Embedded Systems Design

By the same author

VMEbus: a practical companion

Newnes UNIX™ Pocket Book

Microprocessor architectures: RISC, CISC and DSP

Effective PC networking

PowerPC: a practical companion

The PowerPC Programming Pocket Book

The PC and MAC handbook

The Newnes Windows NT Pocket Book

Multimedia Communications

Essential Linux

Migrating to Windows NT

All books published by Butterworth-Heinemann

About the author:

Through his work with Motorola Semiconductors, the author has been involved in the design and development of microprocessor-based systems since 1982. These designs have included VMEbus systems, microcontrollers, IBM PCs, Apple Macintoshes, and both CISC- and RISC-based multiprocessor systems, while using operating systems as varied as MS-DOS, UNIX, Macintosh OS and real-time kernels.

An avid user of computer systems, he has had over 60 articles and papers published in the electronics press, as well as several books.

Embedded Systems Design

Steve Heath

Newnes

An imprint of Butterworth-Heinemann

Newnes
An imprint of Butterworth-Heinemann
Linacre House, Jordan Hill, Oxford OX2 8DP
225 Wildwood Avenue, Woburn, MA 01801–2041
A division of Reed Educational and Professional Publishing Ltd

℞ A member of the Reed Elsevier plc group

OXFORD   AUCKLAND   BOSTON
JOHANNESBURG   MELBOURNE   NEW DELHI

First published 1997
Reprinted 1998, 1999

British Library Cataloguing in Publication Data
A catalogue record for this book is available from the British Library

ISBN 0 7506 3237 2

Typeset by *Steve Heath*

Printed and bound in Great Britain by
Biddles Ltd, Guildford and King's Lynn

Contents

Preface

The term embedded system design covers a very wide range of microprocessor designs and does not simply start and end with a simple microcontroller. It can be a PC running software other than Windows and word processing software. It can be a sophisticated multiprocessor design using the fastest processors on the market today.

The common thread to embedded system design is an understanding of the interaction that the various components within the system have with each other. It is important to understand how the hardware works and the restraints that using a certain peripheral may have on the rest of the system. It is essential to know how to develop the software for such systems and the effect that different hardware designs can have on the software and vice versa. It is this system design knowledge that has been captured in this book as a series of tutorials on the various aspects of embedded system design.

Chapter 1 defines what is meant by the term and in essence defines the scope of the rest of the book. The second chapter provides a set of tutorials on processor architectures explaining the different philosophies that were used in their design and creation. It covers many of the common processor architectures ranging from 8 bit microcontrollers through CISC and RISC processors and finally ending with digital signal processors.

The third chapter discusses different memory types and their uses. The next chapter goes through basic peripherals such as parallel and serial ports along with timers and DMA controllers. This theme is continued in the following chapter which covers analogue to digital conversion and basic power control.

Interrupts are covered in great detail in the sixth chapter because they are so essential to any embedded design. The different types that are available and their associated software routines are described with several examples of how to use them and, perhaps more importantly, how not to use them.

The theme of software is continued in the next two chapters which cover real-time operating systems and software development. Again, these have a tremendous effect on embedded designs but whose design implications are often not well understood or explained. Chapter 9 discusses debugging and emulation techniques.

The remaining four chapters are dedicated to design examples covering buffer and data structures, memory and processor performance trade-offs and techniques, software design examples including using a real-time operating system to create state machines and finally a couple of design examples.

Finally, I would like to thank Duncan Enright at Butterworth-Heinemann for his help, patience, encouragement and support. Special thanks must again go to Sue Carter for yet more editing, intelligent criticism and delicious banana-carrot cake when I needed it.

Steve Heath

Acknowledgements

By the nature of this book, many hardware and software products are identified by their tradenames. In these cases, these designations are claimed as legally protected trademarks by the companies that make these products. It is not the author's nor the publisher's intention to use these names generically, and the reader is cautioned to investigate a trademark before using it as a generic term, rather than a reference to a specific product to which it is attached. The following trademarks mentioned within the text are acknowledged:

- MC68000, MC68020, MC68030, MC68040 are all trademarks of Motorola, Inc.
- PowerPC is a trademark of IBM.
- iAPX8086, iAPX80286, iAPX80386, iAPX80486 and Pentium are trademarks of Intel Corporation.

Many of the techniques within this book can destroy data and such techniques must be used with extreme caution. Again, neither author nor publisher assume any responsibility or liability for their use or any results.

While the information contained in this book has been carefully checked for accuracy, the author assumes no responsibility or liability for its use, or any infringement of patents or other rights of third parties which would result.

As technical characteristics are subject to rapid change, the data contained are presented for guidance and education only. For exact detail, consult the relevant standard or manufacturers' data and specification.

1   What is an embedded system?

When ever the word microprocessor is mentioned, it conjures up a picture of a desktop or laptop PC running an application such as a word processor or a spread sheet. While this is a popular application for microprocessors, it is not the only one and the fact is most people use microprocessors indirectly in common objects and appliances without realising it. Without the microprocessor, they would not be as sophisticated or cheap as they are today.

The embedding of microprocessors into equipment and consumer appliances started before the appearance of the PC and consumes the majority of microprocessors that are made today. In this way, embedded microprocessors are more deeply ingrained into everyday life than any other electronic circuit that is made. A large car may have over 50 microprocessors controlling functions such as the engine through engine management systems, brakes with electronic anti-lock brakes, transmission with traction control and electronically controlled gearboxes, safety with airbag systems, electric windows, air-conditioning and so on. With a well-equipped car, nearly every aspect has some form of electronic control associated with it and thus a need for a microprocessor within an embedded system.

A washing machine may have a microcontroller that contains the different washing programs, provides the power control for the various motors and pumps and even controls the display that tells you how the wash cycles are proceeding.

Mobile phones contain more processing power than a desktop processor of a few years ago. Many toys contain microprocessors and there are even kitchen appliances such as washing machines that use microprocessor-based control systems. The word control is very apt for embedded systems because in virtually every embedded system application, the goal is to control an aspect of a physical system such as temperature, motion, and so on using a variety of inputs. As a result, the skills behind embedded system design are as diverse as the systems that have been built although they share a common heritage.

What is an embedded system?

There are many definitions for this but the best way to define it is to describe it in terms of what it is not and with examples of how it is used.

An embedded system is a microprocessor-based system that is built to control a function or range of functions and is not designed to be programmed by the end user in the same way that a PC is. Yes, a user can make choices concerning functionality but cannot change the functionality of the system by adding/replacing software. With a PC, this is exactly what a user can do: one minute the PC is a word processor and the next it's a games machine simply by changing the software. An embedded system is designed to perform one particular task albeit with choices and different options. The last point is important because it differentiates itself from the world of the PC where the end user does reprogram it whenever a different software package is bought and run. If this need to control the physical world is so great, what is so special about embedded systems that has led to the widespread use of microprocessors? There are several major reasons and these have increased over the years as the technology has progressed and developed.

Replacement for discrete logic-based circuits

The microprocessor came about almost by accident as a programmable replacement for calculator chips in the late 1970s. Up to this point, most control systems using digital logic were implemented using individual logic integrated circuits to create the design and as more functionality became available, the number of chips was reduced.

This was the original reason for a replacement for digital systems constructed from logic circuits. The microprocessor was originally developed to replace a mass of logic that was used to create the first electronic calculators in the early 1970s. For example, the early calculators were made from discrete logic chips and many hundreds were needed just to create a simple four function calculator. As the integrated circuit developed, the individual logic functions were integrated to create higher level functions. Instead of creating an adder from individual logic gates, a complete adder could be bought in one package. It was not long before complete calculators were integrated onto a single chip. This enabled them to be built at a very low cost compared to the original machines but any changes or improvements required that a new chip be developed. The answer was to build a chip that had some form of programmable capability within it. Why not build a chip that took data in, processed it and sent it out again? In this way, instead of creating new functions by analysing the gate level logic and modifying it — a very time-consuming process — new products could be created by changing the program code that processed the information. Thus the microprocessor was born.

Provide functional upgrades

In the same way that the need to develop new calculator chips faster and with less cost prompted the development of the first microprocessors, the need to add or remove functionality from embedded system designs is even more important. With much of the system's functionality encapsulated in the software that runs in the system, it is possible to change and upgrade systems by changing the software while keeping the hardware the same. This reduces the cost of production lower because many different systems can share the same hardware base.

In some cases, this process is not possible or worthwhile but allows the manufacturer to develop new products far quicker and faster. Examples of this include timers and control panels for domestic appliances such as VCRs and televisions.

In other cases, the system can be upgraded to improve functionality. This is frequently done with machine tools, telephone switchboards and so on. The key here is that the ability to add functionality now no longer depends on changing the hardware but can be done by simply changing the software. If the system is connected to a communications link such as a telephone or PC network, then the upgrade can be done remotely without having to physically send out an engineer or technician.

Provide easy maintenance upgrades

The same mechanism that allows new functionality to be added through reprogramming also is beneficial in allowing bugs to be solved through changing software. Again it can reduce the need for expensive repairs and modifications to the hardware.

Improves mechanical performance

For any electromechanical system, the ability to offer a finer degree of control is important. It can prevent excessive mechanical wear, better control and diagnostics and, in some cases, actually compensate for mechanical wear and tear. A good example of this is the engine management system. Here, an embedded microprocessor controls the fuel mixture and ignition for the engine and will alter the parameters and timing depending on inputs from the engine such as temperature, the accelerator position and so on. In this way, the engine is controlled far more efficiently and can be configured for different environments like power, torque, fuel efficiency and so on. As the engine components wear, it can even adjust the parameters to compensate accordingly or if they are dramatically out of spec, flag up the error to the driver or indicate that servicing is needed.

This level of control is demonstrated by the market in 'chipped' engine management units where third party companies modify the software within the control unit to provide more power or torque. The differences can range from 10% to nearly 50% for some turbo charged engines! All this from simply changing a few bytes. Needless to say, this practice may invalidate any guarantee from the manufacturer and may unduly stress and limit the engine's mechanical life. In some cases, it may even infringe the original manufacturer's intellectual property rights.

Protection of intellectual property

To retain a competitive edge, it is important to keep the design knowledge within the company and prevent others from understanding exactly what makes a product function. This knowledge, often referred to as IPR (intellectual property rights), becomes all important as markets become more competitive. With a design that is completely hardware based and built from off-the-shelf components, it can be difficult to protect the IPR that was used in its design. All that is needed to do is to take the product, identify the chips and how they are connected by tracing the tracks on the circuit board. Some companies actually grind the part numbers off the integrated circuits to make it harder to reverse engineer in this way.

With an embedded system, the hardware can be identified but the software that really supplies the system's functionality can be hidden and more difficult to analyse. With self-contained microcontrollers, all that is visible is a plastic package with a few connections to the outside world. The software is already burnt into the on-chip memory and is effectively impossible to access. As a result, the IPR is much more secure and protected.

Replacement for analogue circuits

The movement away from the analogue domain towards digital processing has gathered pace recently with the advent of high performance and low cost processing.

To understand the advantages behind digital signal processing, consider a simple analogue filter. The analogue implementation is extremely simple compared to its digital equivalent. The analogue filter works by varying the gain of the operational amplifier which is determined by the relationship between r_i and r_f.

In a system with no frequency component, the capacitor c_i plays no part as its impedance is far greater than that of r_f. As the frequency component increases, the capacitor impedance decreases until it is about equal with r_f where the effect will be to reduce the gain of the system. As a result, the amplifier acts as a low pass filter where high frequencies will be filtered out. The equation shows the relationship where $j\omega$ is the frequency component. These filters are easy to design and are cheap to build. By making the CR network more complex, different filters can be designed.

The digital equivalent is more complex requiring several electronic stages to convert the data, process it and reconstitute the data. The equation appears to be more involved, comprising of a summation of a range of calculations using sample data multiplied by a constant term. These constants take the place of the CR components in the analogue system and will define the filter's transfer function. With digital designs, it is the tables of coefficients that are dynamically modified to create the different filter characteristics.

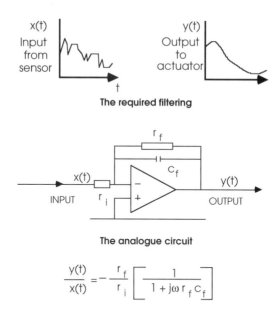

The required filtering

The analogue circuit

$$\frac{y(t)}{x(t)} = -\frac{r_f}{r_i}\left[\frac{1}{1 + j\omega\, r_f\, C_f}\right]$$

The mathematical function

Analogue signal processing

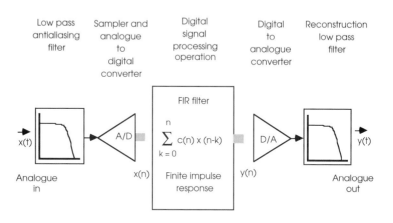

Digital signal processing

Given the complexity of digital processing, why then use it? The advantages are many. Digital processing does not suffer from component ageing, drift or any adjustments which can plague an analogue design. They have high noise immunity and power supply rejection and due to the embedded processor can easily provide self-test features. The ability to dynamically modify the coefficients and therefore the filter characteristics allows complex filters and other functions to be easily implemented. However, the processing power needed to complete the 'multiply–accumulate' processing of the data does pose some interesting processing requirements. The diagram shows the problem. An analogue signal is sampled at a frequency f$_S$ and is converted by the A/D converter. This frequency will be first determined by the speed of this conversion. Every period, t$_S$, there will be a new sample to process using N instructions. The table shows the relationship between sampling speed, the number of instructions and the instruction execution time. It shows that the faster the sampling frequency, the more processing power is needed. To achieve the 1 MHz frequency, a 10 MIPS processor is needed whose instruction set is powerful enough to complete the processing in under 10 instructions. This analysis does not take into account A/D conversion delays. For DSP algorithms, the sampling speed is usually twice the frequency of the highest frequency signal being processed: in this case the 1 MHz sample rate would be adequate for signals up to 500 kHz.

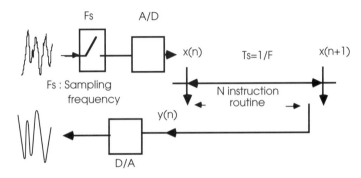

Instruction cycle	Fs	Ts	No. of instructions between two samples
1 µs	1 khz	1 ms	1k
	10 khz	100 µs	100
	100 khz	10 µs	10
	1Mhz	1 µs	1
100 ns	1 khz	1 ms	10k
	10 khz	100 µs	1k
	100 khz	10 µs	100
	1Mhz	1 µs	10

DSP processing requirements

One major difference between analogue and digital filters is the accuracy and resolution that they offer. Analogue signals may have definite limits in their range, but have infinite values between that range. Digital signal processors are forced to represent these infinite variations within a finite number of steps determined by the number of bits in the word. With an 8 bit word, the increases are in steps of 1/256 of the range. With a 16 bit word, such steps are in 1/65536 and so on. Depicted graphically as shown, a 16 bit word would enable a low pass filter with a roll-off of about 90 dB. A 24 bit word would allow about 120 dB roll-off to be achieved.

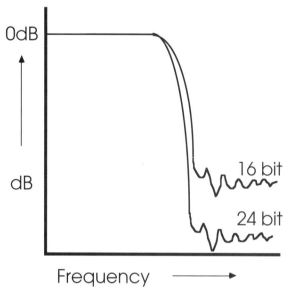

Word size and cutoff frequencies

DSP can be performed by ordinary microprocessors, although their more general-purpose nature often limits performance and the frequency response. However, with responses of only a few hundred Hertz, even simple microcontrollers can perform such tasks. As silicon technology improved, special building blocks appeared allowing digital signal processors to be developed, but their implementation was often geared to a hardware approach rather than designing a specific processor architecture for the job.

Inside the embedded system

Processor

The main criteria for the processor is: can it provide the processing power needed to perform the tasks within the system? This seems obvious but it frequently occurs that the tasks are either underestimated in terms of their size and/or complexity or that creeping elegance expands the specification to beyond the processor's capability.

In many cases, these types of problems are compounded by the performance measurement used to judge the processor. Benchmarks may not be representative of the type of work that the system is doing.

They may execute completely out of cache memory and thus give an artificially high performance level which the final system cannot meet because its software does not fit in the cache. The software overheads for high level languages, operating systems and interrupts may be higher than expected. These are all issues that can turn a paper design into failed reality.

While processor performance is essential and forms the first gating criterion, there are others such as cost — this should be system cost and not just the cost of the processor in isolation, power consumption, software tools and component availability and so on.

These topics are discussed in more detail in Chapter 2.

Memory

Memory is an important part of any embedded system design and is heavily influenced by the software design, and in turn may dictate how the software is designed, written and developed. These topics will be addressed in more detail later on in this book. As a way of introduction, memory essentially performs two functions within an embedded system:

- It provides storage for the software that it will run

  At a minimum, this will take the form of some non-volatile memory that retains its contents when power is removed. This can be on-chip read only memory (ROM) or external EPROM. The software that it contains might be the complete program or an initialisation routine that obtains the full software from another source within or outside of the system. This initialisation routine is often referred to as a bootstrap program or routine. PC boards that have embedded processors will often start up using software stored in an onboard EPROM and then wait for the full software to be downloaded from the PC across the PC expansion bus.

- It provides storage for data such as program variables and intermediate results, status information and any other data that might be created throughout the operation

  Software needs some memory to store variables and to manage software structures such as stacks. The amount of memory that is needed for variables is frequently less than that needed for the actual program. With RAM being more expensive than ROM and non-volatile, many embedded systems and in particular, microcontrollers, have small amounts of RAM compared to the ROM that is available for the program. As a result, the software that is written for such systems often has to be written to minimise RAM usage so that it will fit within the memory resources placed upon the design. This will often mean the use of compilers that produce ROMable code that does not rely on being resident in RAM to execute. This is discussed in more detail in Chapter 3.

Peripherals

An embedded system has to communicate with the outside world and this is done by peripherals. Input peripherals are usually associated with sensors that measure the external environment and thus effectively control the output operations that the embedded system performs. In this way, an embedded system can be modelled on a three-stage pipeline where data and information input into the first stage of the pipeline, the second stage processes it before the third stage outputs data.

If this model is then applied to a motor controller, the inputs would be the motor's actual speed and power consumption, and the speed required by the operator. The outputs would be a pulse width modulated waveform that controls the power to the motor and hence the speed and an output to a control panel showing the current speed. The middle stage would be the software that processed the inputs and adjusts the outputs to achieve the required engine speed. The main types of peripherals that are used include:

*   Binary outputs

    These are simple external pins whose logic state can be controlled by the processor to either be a logic zero (off) or a logic one (on). They can be used individually or grouped together to create parallel ports where a group of bits can be input or output simultaneously.

*   Serial outputs

    These are interfaces that send or receive data using one or two pins in a serial mode. They are less complex to connect but are more complicated to program. A parallel port looks very similar to a memory location and is easier to visualise and thus use. A serial port has to have data loaded into a register and then a start command issued. The data may also be augmented with additional information as required by the protocol.

*   Analogue values

    While processors operate in the digital domain, the natural world does not and tends to orientate to analogue values. As a result, interfaces between the system and the external environment need to be converted from analogue to digital and vice versa.

*   Displays

    Displays are becoming important and can vary from simple LEDs and seven segment displays to small alpha-numeric LCD panels.

*   Time derived outputs

    Timers and counters are probably the most commonly used functions within an embedded system.

Software

The software components within an embedded system often encompasses the technology that adds value to the system and defines what it does and how well it does it. The software can consist of several different components:

- Initialisation and configuration
- Operating system or run time environment
- The applications software itself
- Error handling
- Debug and maintenance support

Algorithms

Algorithms are the key constituents of the software that makes an embedded system behave in the way that it does. They can range from mathematical processing through to models of the external environment which are used to interpret information from external sensors and thus generate control signals.

Defining and implementing the correct algorithm is a critical operation and is described in detail in the chapters on software development.

Examples

This section will go through some example embedded systems and briefly outline the type of functionality that each offers.

Microcontroller

Microcontrollers can be considered as self-contained systems with a processor, memory and peripherals so that in many cases all that is needed to use them within an embedded system is to add software. The processors are usually based on 8 bit stack-based architectures such as the MC6800 family. There are 4 bit versions available such as the National COP series which further reduce the processing power to reduce cost even further.

Microcontrollers are usually available in several forms:

- Devices for prototyping or low volume production runs

    These devices use non-volatile memory to allow the software to be downloaded and returned in the device. UV erasable EPROM used to be the favourite but EEPROM is also gaining favour. Some microcontrollers used a special package with a piggyback socket on top of the package to allow an external EPROM to be plugged in for prototyping. This memory technology replaces the ROM on the chip allowing software to be downloaded and debugged. The device can be reprogrammed as needed until the software reaches its final release version.

The use of non-volatile memory also makes these devices suitable for low volume production runs or where the software may need customisation and thus preventing moving to a ROMed version.

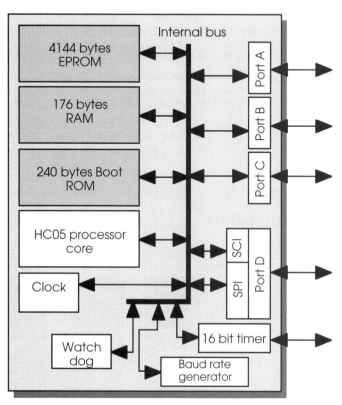

Example microcontroller (Motorola MC68HC705C4A)

These devices are sometimes referred to as umbrella devices with a single device capable of providing prototyping support for a range of other controllers in the family.

- Devices for low to medium volume production runs

   In the mid-1980s, a derivative of the prototype device appeared on the market called the one time programmable or OTP. These devices use EPROM instead of the ROM but instead of using the ceramic package with a window to allow the device to be erased, it was packaged in a cheaper plastic pack and thus was only capable of programming a single time — hence the name. These devices are cheaper than the prototype versions but still have the disadvantage of programming. However, their lower cost has made them a suitable alternative to producing a ROM device. For production quantities in the low to medium range, they are cost effective and offer the ability to customise soft-ware as necessary.

- Devices for high volume production runs

For high volumes, microcontrollers can be built already pro-grammed with software in the ROM. To do this a customer supplies the software to the manufacturer who then creates the masks necessary to create the ROM in the device. This process is normally done on partly processed silicon wafers to reduce the turnaround time. The advantage for the customer is that the costs are much lower than using prototyping or OTP parts and there is no programming time or overhead involved. The downside is that there is usually a minimum order based on the number of chips that a wafer batch can produce and an upfront mask charge. The other major point is that once in ROM, the software cannot be changed and therefore customisation or bug fixing would have to wait until the next order or involve scrapping all the devices that have been made. It is possible to offer some customisation by including different software mod-ules and selecting the required ones on the basis of a value read into the device from an external port but this does consume memory which can increase the costs.

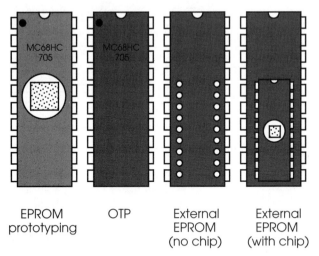

| EPROM prototyping | OTP | External EPROM (no chip) | External EPROM (with chip) |

Prototype microcontrollers

Expanded microcontroller

The choice of memory sizes and partitioning is usually a major consideration. Some applications require more memory or peripher-als than are available on a standard part. Most microcontroller families have parts that support external expansion and have an external memory and/or I/O bus which can allow the designer to put almost any configuration together. This is often done by using a parallel port as the interface instead of general-purpose I/O. Many of the higher performant microcontrollers are adopting this approach.

An expanded microcontroller

In the example shown, the microcontroller has an expanded mode that allows the parallel ports A and B to be used as byte wide interfaces to external RAM and ROM. In this type of configuration, some microcontrollers disable access to the internal memory while others still allow it.

Microprocessor based

Microprocessor-based embedded systems originally took existing general-purpose processors such as the MC6800 and 8080 devices and constructed systems around them using external peripherals and memory. The use of processors in the PC market continued to provide a series of faster and faster processors such as the MC68020, MC68030 and MC68040 devices from Motorola and the 80286, 80386, 80486 and Pentium devices from Intel. These CISC architectures have been complemented with RISC processors such as the PowerPC, MIPS and others. These systems offer more performance than is usually available from a traditional microcontroller.

However, this is beginning to change. There has been the development of integrated microprocessors where the processor is combined with peripherals such as parallel and serial ports, DMA controllers and interface logic to create devices that are more suitable for embedded systems by reducing the hardware design task and costs. As a result, there has been almost a parallel development of these integrated processors along with the desktop processors. Typically, the integrated processor will use a processor generation that is one behind the current generation. The reason is dependent on silicon technology and cost. By using the previous generation which is smaller, it frees up silicon area on the die to add the peripherals and so on.

Board based

So far, the types of embedded systems that we have considered have assumed that the hardware needs to be designed, built and debugged. An alternative is to use hardware that has already been built and tested such as board-based systems as provided by PCs and through international board standards as VMEbus. The main advantage is the reduced work load and the availability of ported software that can simply be utilised with very little effort. The disadvantages are higher cost and in some cases restrictions in the functionality that is available.

2 Embedded processors

The development of processors for embedded system design has essentially followed the development of microprocessors as a whole. The processor development has provided the processing heart for architecture which combined with the right software and hardware peripherals has become an embedded design. With the advent of better fabrication technology supporting higher transistor counts and lower power dissipation, the processor core has been integrated with peripherals and memory to provide standalone microcontrollers or integrated processors that only need the addition of external memory to provide a complete hardware system suitable for embedded design. The scope of this chapter is to explain the strengths and weaknesses of various architectures to provide a good understanding of the trade-offs involved in choosing and exploiting a processor family.

There are essentially four basic architecture types which are usually defined as 8 bit accumulator, 16/32 bit complex instruction set computers (CISC), reduced instruction set computer (RISC) architectures and digital signal processors (DSP). Their development or to be more accurate, their availability to embedded system designers is chronological and tends to follow the same type of pattern as shown in the graph.

	1975	1980	1984	1989	1993
Highest performance	MC6800	MC68000	MC68020	MC68040	MC68060
Medium performance		MC6800	MC68000	MC68020	MC68040
Lowest performance			MC6800	MC68000	MC68020
Cost-effective performance				MC6800	MC68000
End of life					MC6800

Processor life history

However, it should be remembered that in parallel with this life cycle, processor architectures are being moved into microcontroller and integrated processor devices so that the end of life really refers to the discontinuance of the architecture as a separate CPU plus external memory and peripherals product. The MC6800 processor is no longer used in discrete designs but there are over 200 MC6801/6805 and 68HC11 derivatives that essentially use the same basic architecture and instruction set.

8 bit accumulator processors

This category of processor first appeared in the mid-1970s as the first microprocessors. Devices such as the 8080 from Intel and the MC6800 from Motorola started the microprocessor revolution. They provided about 1 MIP of performance and were at their introduction the fastest processors available.

Register models

The programmer has a very simple register model for this type of processor. The model for the Motorola MC6800 8 bit processor is shown as an example. It has two 8 bit accumulators used for storing data and performing arithmetic operations. The program counter is 16 bits in size and two further 16 bit registers are provided for stack manipulations and address indexing.

```
    15                    7                    0

                              | Accumulator A |

                              | Accumulator B |

                        |     Index register X     |

                        |     Program counter      |

                        |      Stack pointer       |

                              | Condition code |
```

The MC6800 programmer's model

On first inspection, the model seems quite primitive and not capable of providing the basis of a computer system. There do not seem to be enough registers to hold data, let alone manipulate it! What is often forgotten is that many of the instructions, such as logical operations, can operate on direct memory using the index register to act as pointer. This removes the need to bring data into the processor at the expense of extra memory cycles and the need for additional or wider registers. The main area within memory that is used for data storage is known as the stack. It is normally accessed using a special register that indexes into the area called the stack pointer. This is used to provide local data storage for programs and to store information for the processor such as return addresses for subroutine jumps and interrupts.

The stack pointer provides additional storage for the programmer: it is used to store data like return addresses for subroutine calls and provides additional variable storage using a PUSH/POP mechanism. Data is PUSHed onto the stack to store it, and POPed off to retrieve it. Providing the programmer can track where the data resides in these stack frames, it offers a good replacement for the missing registers.

8 bit data restrictions

An 8 bit data value can provide an unsigned resolution of only 256 bits, which makes it unsuitable for applications where a higher resolution is needed. In these cases, such as financial, arithmetic, high precision servo control systems, the obvious solution is to increase the data size to 16 bits. This would give a resolution of 65536 — an obvious improvement. This may be acceptable for a control system but is still not good enough for a data processing program, where a 32 bit data value may have to be defined to provide sufficient integer range. While there is no difficulty with storing 8, 16, 32 or even 64 bits in external memory, even though this requires multiple bus accesses, it does prevent the direct manipulation of data through the instruction set.

However, due to the register model, data larger than 8 bits cannot use the standard arithmetic instructions applicable to 8 bit data stored in the accumulator. This means that even a simple 16 bit addition or multiplication has to be carried out as a series of instructions using the 8 bit model. This reduces the overall efficiency of the architecture.

The code example is a routine for performing a simple 16 bit multiplication. It takes two unsigned 16 bit numbers and produces a 16 bit product. If the product is larger than 16 bits, only the least significant 16 bits are retained. The first eight or so instructions simply create a temporary storage area on the stack for the multiplicand, multiplier, return address and loop counter. Compared to internal register storage, storing data in stack frames is not as efficient due the increased external memory access.

Accessing external data consumes machine cycles which could be used to process data. Without suitable registers and the 16 bit wide accumulator, all this information must be stored externally on the stack. The algorithm used simply performs a succession of arithmetic shifts on each half of the multiplicand stored in the A and B accumulators. Once this is complete, the 16 bit result is split between the two accumulators and the temporary storage cleared off the stack. The operation takes at least 29 instructions to perform with the actual execution time totally dependant on the values being multiplied together. For comparison, most 16/32 bit processors such as the MC68000 and 80x86 families can perform the same operation with a single instruction!

```
MULT16  LDX    #5              CLEAR WORKING REGISTERS
        CLR    A
LP1            STA    A    U-1,X
        DEX
        BNE           LP1
        LDX    #16             INITIAL SHIFT COUNTER
LP2     LDA    A    Y+1           GET Y(LSBIT)
        AND    A    #1
        TAB                  SAVE Y(LSBIT) IN ACCB
        EOR    A    FF      CHECK TO SEE IF YOU ADD
        BEQ    SHIFT         OR SUBTRACT
        TST    B
        BEQ           ADD
        LDA    A    U+1
        LDA    B    U
        SUB    A    XX+1
        SBC    B    XX
        STA    A    U+1
        STA    B    U
        BRA    SHIFT         NOW GOTO SHIFT ROUTINE
ADD            LDA    A    U+1
        LDA    B    U
        ADD    A    XX+1
        ADC    B    XX
        STA    A    U+1
        STA    B    U
SHIFT          CLR    FF      SHIFT ROUTINE
        ROR         Y
        ROR         Y+1
        ROL         FF
        ASR         U
        ROR         U+1
        ROR         U+2
        ROR         U+3
        DEX
        BNE         LP2
        RTS                    FINISH SUBROUTINE
        END
```

M6800 code for a 16 bit by 16 bit multiply

Addressing memory

When the first 8 bit microprocessors appeared during the middle to late 1970s, memory was expensive and only available in very small sizes: 256 bytes up to 1 kilobyte. Applications were small, partly due to their implementation in assembler rather than a high level language, and therefore the addressing range of 64 kilobytes offered by the 16 bit address seemed extraordinarily large. It was unlikely to be exceeded. As the use of these early microprocessors became more widespread, applications started to grow in size and the use of operating systems like CP/M and high level languages increased memory requirements until the address range started to limit applications. Various techniques like bank switching and program overlays were developed to help.

System integrity

Another disadvantage with this type of architecture is its unpredictability in handling error conditions. A bug in a software application could corrupt the whole system, causing a system to either crash, hang up or, even worse, perform some unforeseen operations. The reasons are quite simple: there is no partitioning between data and programs within the architecture. An application can update a data structure using a corrupt index pointer which overwrites a part of its program.

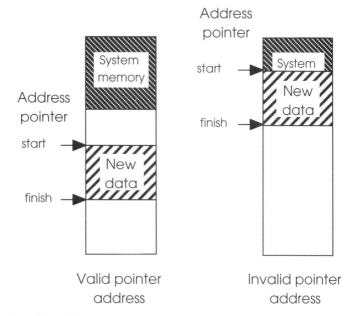

System corruption via an invalid pointer

Data are simply bytes of information which can be interpreted as instruction codes. The processor calls a subroutine within this area, starts to execute the data as code and suddenly the whole system starts performing erratically! On some machines, certain undocumented code sequences could put the processor in a test mode and start cycling through the address ranges etc. These attributes restricted their use to non-critical applications.

Example 8 bit architectures

Z80

The Z80 microprocessor is an 8 bit CPU with a 16 bit address bus capable of direct access of 64k of memory space. It was designed by Zilog and rapidly gained a lot of interest. The Z80 was based on the Intel 8080 but has an extended instruction set and many hardware improvements. It can run 8080 code if needed by its support of the 8080 instruction set. The instruction set is essential based around an 8 bit op code giving a maximum of 256 instructions. The 158 instructions that are specified — the others are reserved — include 78 instructions from the 8080. The instruction set supports the use of extension bytes to encode additional information. In terms of processing power, it offered about 1 MIP at 4 MHz clock speed with a minimum instruction time of 1 μs and a maximum instruction time of 5.75 μs.

The programming model includes an accumulator and six 8 bit registers that can be paired together to create three 16 bit registers. In addition to the general registers, a stack pointer, program counter, and two index (memory pointers) registers are provided. It uses external RAM for its stack. While not as powerful today as a PowerPC or Pentium, it was in its time a very powerful processor and was used in many of the early home computers such as the Amstrad CPC series. It was also used in many embedded designs partly because of its improved performance and also for its built-in refresh circuitry for DRAMs. This circuitry greatly simplified the external glue logic that was needed with DRAMs.

Pin	Signal	Pin	Signal
1	A11	21	RD
2	A12	22	WR
3	A13	23	BUSAK
4	A14	24	WAIT
5	A15	25	BUSRQ
6	CLOCK	26	RESET
7	D4	27	M1
8	D3	28	RFSH
9	D5	29	GND
10	D6	30	A0
11	Vcc	31	A1
12	D2	32	A2
13	D7	33	A3
14	D0	34	A4
15	D1	35	A5
16	INT	36	A6
17	NMI	37	A7
18	HALT	38	A8
19	MREQ	39	A9
20	IORQ	40	A10

The Z80 signals

The Z80 was originally packaged in a 40 pin DIP package and ran at 2.5 and 4 MHz. Since then other packages and speeds have become available including low power CMOS versions — the original was made in NMOS and dissipated about 1 watt. Zilog now use the processor as a core within its range of Z800 microcontrollers with various configurations of on-chip RAM and EPROM.

Signal	Description
A0 - A15	Address bus output tri-state
D0 - D7	Data bus bidirectional tri-state
CLOCK	CPU clock input
RFSH	Dynamic memory refresh output
HALT	CPU halt status output
RESET	Reset input
INT	Interrupt request input (active low)
NMI	Non-maskable interrupt input (active low)
BUSRQ	Bus request input (active low)
BUSAK	Bus acknowledge output (active low)
WAIT	Wait request input (active low)
RD, WR	Read and write signals
IORQ	I/O operation status output
MREQ	Memory refresh output
M1	Output pulse on instruction fetch cycle
Vcc	+5 volts
GND	0 volts

The Z80 pinout descriptions

Z80 programming model

The Z80 programming model essential consists of a set of 8 bit registers which can be paired together to create 16 bit versions for use as data storage or address pointers. There are two register sets within the model: the main and alternate. Only one set can be used at any one time and the switch and data transfer is performed by the EXX instruction. The registers in the alternate set are designated by a ´ suffix.

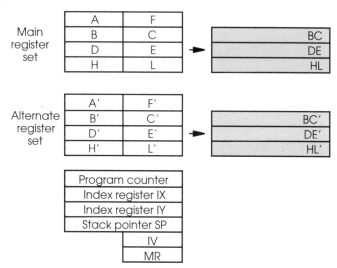

The Z80 programming model

The model has an 8 bit accumulator A and a flags register known as F. This contains the status information such as carry, zero, sign and overflow. This register is also known as PSW (program status word) in some documentation. Registers B, C, D, E, H and L are 8 bit general-purpose registers that can be paired to create 16 registers known as BC, DE and HL. The remaining registers are the program counter PC, two index registers IX and IY and a stack pointer SP. All these four registers are 16 bits in size and can access the whole 64 kbytes of external memory that the Z80 can access. There are two additional registers IV and MR which are the interrupt vector and the memory refresh registers. The IV register is used in the interrupt handling mode 2 to point to the required software routine to process the interrupt. In mode 1, the interrupt vector is supplied via the external data bus. The memory refresh register is used to control the on-chip DRAM refresh circuitry.

Unlike the MC6800, the Z80 does not use memory mapped I/O and instead uses the idea of ports, just like the 8080. The lower 8 bits of the address bus are used along with the IORQ signal to access any external peripherals. The IORQ signal is used to differentiate the access from a normal memory cycle. These I/O accesses are similar from a hardware perspective to a memory cycle but only occur when an I/O port instruction (IN, OUT) is executed. In some respects, this is similar to the RISC idea of load and store instructions to bring information into the processor, process it and then write out the data. This system gives 255 ports and is usually sufficient for most embedded designs.

MC6800

The MC6800 was introduced in the mid-1970s by Motorola and is as an architecture the basis of several hundred derivative processors and microcontrollers such as the MC6809, MC6801, MC68HC05, MC68HC11, MC68HC08 families.

The processor architecture is 8 bits and uses a 64 kbyte memory map. Its programming model uses two 8 bit accumulators and a single 16 bit index register. Later derivatives such as the MC68HC11 added an additional index register and allowed the two accumulators to be treated as a single 16 bit accumulator to provide additional support for 16 bit arithmetic.

Its external bus was synchronous with separate address and data ports and the device operated at either 1, 1.5 or 2 MHz. The instruction set was essential based around an 8 bit instruction with extensions for immediate values, address offsets and so on. It supported both non-maskable and software interrupts.

These type of processors have largely been replaced today by the microcontroller versions which have the same or advanced processor architectures and instruction sets but have the added advantage of glueless interfaces to memory and peripherals incorporated onto the chip itself. Discrete processors are still used but these tend to be the higher performance devices such as the MC68000 and 80x86 processors. But even with these faster and higher performance devices, the same trend of moving to integrated microcontroller type of devices is being followed as even higher performance processors such as RISC devices become available.

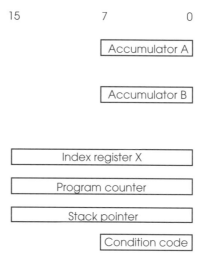

15   7   0

Accumulator A

Accumulator B

Index register X

Program counter

Stack pointer

Condition code

The MC6800 programmer's model

Microcontrollers

The previous section has described the 8 bit processors. While most of the original devices are no longer available, their architectures live on in the form of microcontrollers. These devices do not need much processing power — although this is now undergoing a radical change as will be explained later — but instead have become a complete integrated computer system by integrating the processor, memory and peripherals onto a single chip.

MC68HC05

The MC68HC05 is microcontroller family from Motorola that uses an 8 bit accumulator-based architecture as its processor core. This is very similar to that of the MC6800 except that it only has a single accumulator.

It uses memory mapping to access any on-chip peripherals and has a 13 bit program counter and effectively a 6 bit stack pointer. These reduced size registers — with many other 8 bit processors such as the Z80/8080 or MC6800, they are 16 bits is size — are used to reduce the complexity of the design. The microcontroller uses on-chip memory and therefore its does not make sense to define registers that can address memory that doesn't exist on the chip. The MC68HC05 family is designed for low cost applications where superfluous hardware is removed to reduce the die size, its power consumption and cost. As a result, the stack pointer points to the start of the on-chip RAM and can only use 64 bytes, and the program counter is reduced to 13 bits.

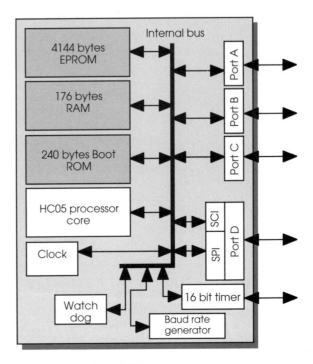

Example microcontroller (Motorola MC68HC705C4A)

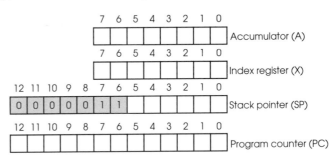

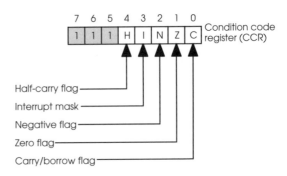

68HC05 programming model

MC68HC11

The 68HC11 is a powerful 8 bit data, 16 bit address microcontroller from Motorola that was on its introduction one of the most powerful and flexible microcontrollers available. It was originally designed in conjunction with General Motors for use within engine management systems. As a result, its initial versions had built-in EEPROM/OTPROM, RAM, digital I/O, timers, 8 channel 8 bit A/D converter, PWM generator, and synchronous and asynchronous communications channels (RS232 and SPI). Its current consumption is low with a typical value of less than 10 mA.

Architecture

The basic processor architecture is similar to that of the 6800 and has two 8 bit accumulators referred to as registers A and B. They can be concatenated to provide a 16 bit double accumulator called register D. In addition, there are two 16 bit index registers X and Y to provide indexing to anywhere within its 64 kbytes memory map.

Through its 16 bit accumulator, the instruction set can support several 16 bit commands such as add, subtract, shift and 16 by 16 division. Multiplies are limited to 8 bit values.

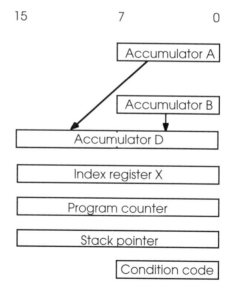

MC68HC11 programming model

Data processors

Processors like the 8080 and the MC6800 provided the computing power for many early desktop computers and their successors have continued to power the desktop PC. As a result, it should not be surprising that they have also provided the processing power for more powerful systems where a microcontroller cannot provide either the processing power or the correct number or type of peripherals. They have also provided the processor cores for more integrated chips which form the next category of embedded systems.

Complex instructions, microcode and nanocode

With the initial development of the microprocessor concentrated on the 8 bit model, it was becoming clear that larger data sizes, address space and more complex instructions were needed. The larger data size was needed to help support higher precision arithmetic. The increased address space was needed to support bigger blocks of memory for larger programs. The complex instruction was needed to help reduce the amount of memory required to store the program by increasing the instruction efficiency: the more complex the instruction, the less needed for a particular function and therefore the less memory that the system needed. It should be remembered that it was not until recently that memory has become so cheap.

The instruction format consists of an op code followed by a source effective address and a destination effective address. To provide sufficient coding bits, the op code is 16 bits in size with further 16 bit operand extensions for offsets and absolute addresses. Internally, the instruction does not operate directly on the internal resources, but is decoded to a sequence of microcode instructions, which in turn calls a sequence of nanocode commands which controls the sequencers and arithmetic logic units (ALU). This is analogous to the many macro subroutines used by assembler programmers to provide higher level 'pseudo' instructions. On the MC68000, microcoding and nanocoding allow instructions to share common lower level routines, thus reducing the hardware needed and allowing full testing and emulation prior to fabrication. Neither the microcode nor the nanocode sequences are available to the programmer.

These sequences, together with the sophisticated address calculations necessary for some modes, often take more clock cycles than are consumed in fetching instructions and their associated operands from external memory. This multi-level decoding automatically lends itself to a pipelined approach which also allows a prefetch mechanism to be employed.

Pipelining works by splitting the instruction fetch, decode and execution into independent stages: as an instruction goes through each stage, the next instruction follows it without waiting for it to completely finish. If the instruction fetch is included within the pipeline, the next instruction can be read from memory, while the preceding instruction is still being executed as shown.

The only disadvantage with pipelining concerns pipeline stalls. These are caused when any stage within the pipeline cannot complete its allotted task at the same time as its peers. This can occur when wait states are inserted into external memory accesses, instructions use iterative techniques or there is a change in program flow.

With iterative delays, commonly used in multiply and divide instructions and complex address calculations, the only possible solutions are to provide additional hardware support, add more stages to the pipeline, or simply suffer the delays on the grounds that the performance is still better than anything else! Additional hardware support may or may not be within a designer's real-estate budget (real estate refers to the silicon die area, and directly the number of transistors available). Adding stages also consumes real estate and increases pipeline stall delays when branching.

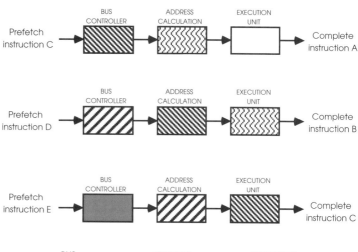

Pipelining instructions

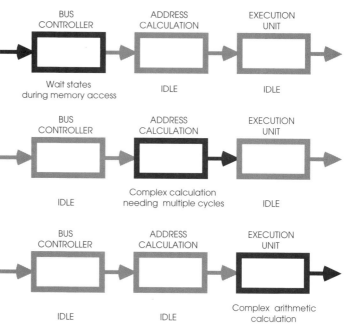

Pipeline stalls

The main culprits are program branching and similar operations. The problem is caused by the decision whether to take the branch or not being reached late in the pipeline, i.e. after the next instruction has been prefetched. If the branch is not taken, this instruction is valid and execution can carry on. If the branch is taken, the instruction is not valid and the whole pipeline must be flushed and reloaded. This causes additional memory cycles before the processor can continue. The delay is dependent on the number of stages, hence the potential difficulty in increasing the number of stages to reduce iterative delays. This interrelation of engineering trade-offs is a common theme within microprocessor architectures. Similar problems can occur for any change of flow: they are not limited to just branch instructions and can occur with interrupts, jumps, software interrupts etc. With the large usage of these types of instructions, it is essential to minimise these delays. The longer the pipeline, the greater the potential delay.

The next question was over how to migrate from the existing 8 bit architectures. Two approaches were used: Intel chose the compatibility route and simply extended the 8080 programming model, while Motorola chose to develop a different architecture altogether which would carry it into the 32 bit processor world.

INTEL 80286

The Intel 80286 was the successor to the 8086 and 8088 processors and offered a larger addressing space while still preserving compatibility with its predecessors.

Architecture

The 80286 has two modes of operation known as real mode and protected mode: real mode describes its emulation of the 8086/8088 processor including limiting its external address bus to 20 bits to mimic the 8086/8088 1 Mbyte address space. In its real mode, the 80286 adds some additional registers to allow access to its larger 16 Mbyte external address space, while still preserving its compatibility with the 8086 and 8088 processors.

The register set comprises of four general-purpose 16 bit registers (AX, BX, CX and DX) and four segment address registers (CS, DS, SS and ES) and a 16 bit program counter. The general-purpose registers — AX, BX, CX, and DX — can be accessed as two 8 bit registers by changing the X suffix to either H or L. In this way, each half of register AX can be accessed as AH or AL and so on for the other three registers.

These registers form a set that is the same as that of an 8086. However, when the processor is switched into its protected mode, the register set is expanded and includes two index registers (DI and SI) and a base pointer register. These additions allow the 80286 to support a simple virtual memory scheme.

Within the IBM PC environment, the 8086 and 8088 processors can access beyond the 1 Mbyte address space by using paging and special hardware to simulate the missing address lines. This additional memory is known as expanded.

Accumulator	AX
Base register	BX
Counter reg	CX
Data register	DX
Source index	SI
Destination index	DI
Stack pointer	SP
Base pointer	BP
Code segment	CS
Data segment	DS
Stack segment	SS
Extra segment	ES
Instruction pointer	IP
Status flags	FL

15   0

Intel 80286 processor register set

Interrupt facilities

The 80286 can handle 256 different exceptions and the vectors for these are held in a vector table. The vector table's construction is different depending on the processor's operating mode. In the real mode, each vector table consists of two 16 bit words that contain the interrupt pointer and code segment address so that the associated interrupt routine can be located and executed. In the protected mode of operation each entry is 8 bytes long.

Vector	Function
0	Divide error
1	Debug exception
2	Non-masked interrupt NMI
3	One byte interrupt INT
4	Interrupt on overflow INTO
S	Array bounds check BOUND
6	Invalid opcode
7	Device not available
8	Double fault
9	Coprocessor segment overrun
10	Invalid TSS
11	Segment not present
12	Stack fault
13	General protection fault
14	Page fault
15	Reserved
16	Coprocessor error
17-32	Reserved
33-255	INT n trap instructions

Instruction set

The instruction set for the 80286 follows the same pattern as that for the Intel 8086 and programs written for the 8086 are compatible with the 80286 processor.

80287 floating point support

The 80286 can also be used with the 80287 floating point coprocessor to provide acceleration for floating point calculations. If the device is not present, it is possible to emulate the floating point operations in software, but at a far lower performance.

Feature comparison

Feature	8086	8088	80286
Address bus	20 bit	20 bit	24 bit
Data bus	16 bit	8 bit	16 bit
FPU present	NO	NO	NO
Memory management	NO	NO	YES
Cache on-chip	NO	NO	NO
Branch acceleration	NO	NO	NO
TLB support	NO	NO	NO
Superscalar	NO	NO	NO
Frequency (MHz)	5,8,10	5,8,10	6,8,10,12
Average cycles/Inst.	12	12	4.9
Frequency of FPU	=CPU	=CPU	2/3 CPU
Frequency	3X	3X	2X
Address range	1 Mbytes	1 Mbytes	16 Mbytes
Frequency scalability	NO	NO	NO
Voltage	5 v	5 v	5 v

Intel 8086, 8088 and 80286 processors

INTEL 80386DX

The 80386 processor was introduced in 1987 as the first 32 bit member of the family. It has 32 bit registers and both 32 bit data and address buses. It is software compatible with the previous generations through the preservation of the older register set within the 80386's newer extended register model and through a special 8086 emulation mode where the 80386 behaves like a very fast 8086.

The processor has an on-chip paging memory management unit which can be used to support multitasking and demand paging virtual memory schemes if required.

Architecture

The 80386 has eight general-purpose 32 bit registers EAX, EBX, ECX, EDX, ESI, EDI, EBP and ESP. These general-purpose registers are used for storing either data or addresses. To ensure compatibility with the earlier 8086 processor, the lower half of each register can be accessed as a 16-bit register (AX, BX, CX, DX, SI, DI, BP and SP). The AX, BX, CX and DX registers can be also accessed as 8 bit registers by changing the X suffix for either H or L thus creating the 8088 registers AH, AL, BH, BL and so on.

To generate a 32 bit physical address, six segment registers (CS, SS, DS, ES, FS, GS) are used with addresses from the general registers or instruction pointer. The code segment (CS) is used with the instruction pointer to create the addresses used for instruction fetches and any stack access uses the SS register. The remaining segment registers are used for data addresses.

Each segment register has an associated descriptor register which is used to program and control the on-chip memory management unit. These descriptor registers — controlled by the operating system and not normally accessible to the application programmer — hold the base address, segment limit and various attribute bits that describe the segment's properties.

The 80386 can run in three different modes: the real mode, where the size of each segment is limited to 64 kbytes, just like the 8088 and 8086; a protected mode, where the largest segment size is increased to 4 Gbytes; and a special version of the protected mode that creates multiple virtual 8086 processor environments.

The 32 bit flag register contains the normal carry zero, auxiliary carry, parity, sign and overflow flags. The resume flag is used with the trap 1 flag during debug operations to stop and start the processor. The remaining flags are used for system control to select virtual mode, nested task operation and input/output privilege level.

Intel 80386 register set

For external input and output, a separate peripheral address facility is available similar to that found on the 8086. As an alternative, memory mapping is also supported (like the M68000 family) where the peripheral is located within the main memory map.

Interrupt facilities

The 80386 has two external interrupt signals which can be used to allow external devices to interrupt the processor. The INTR input generates a maskable interrupt while the NMI generates a non-maskable interrupt and naturally has the higher priority of the two.

During an interrupt cycle, the processor carries out two interrupt acknowledge bus cycles and reads an 8 bit vector number on D0–D7 during the second cycle. This vector number is then used to locate within the vector table the address of the corresponding interrupt service routine. The NMI interrupt is automatically assigned the vector number of 2.

Software interrupts can be generated by executing the INT *n* instruction where *n* is the vector number for the interrupt. The vector table consists of 4 byte entries for each vector and starts at memory location 0 when the processor is running in the real mode. In the protected mode, each vector is 8 bytes long. The vector table is very similar to that of the 80286.

Vector	Function
0	Divide error
1	Debug exception
2	Non-masked interrupt NMI
3	One byte interrupt INT
4	Interrupt on overflow INTO
S	Array bounds check BOUND
6	Invalid opcode
7	Device not available
8	Double fault
9	Coprocessor segment overrun
10	Invalid TSS
11	Segment not present
12	Stack fault
13	General protection fault
14	Page fault
15	Reserved
16	Coprocessor error
17-32	Reserved
33-255	INT n trap instructions

Instruction set

The 80386 instruction set is essentially a superset of the 8086 instruction set. The format follows the dyadic approach and uses two operands as sources with one of them also duplicating as a destination. Arithmetic and other similar operations thus follow the A+B=B type of format (like the M68000). When the processor is operating in the real mode — like an 8086 processor — its instruction set, data types and register model is essentially restricted to a that of the 8086.

In its protected mode, the full 80386 instruction set, data types and register model becomes available. Supported data types include bits, bit fields, bytes, words (16 bits), long words (32 bits) and quad words (64 bits). Data can be signed or unsigned binary, packed or unpacked BCD, character bytes and strings.

In addition, there is a further group of instructions that can be used when the CPU is running in protected mode only. They provide access to the memory management and control registers. Typically, they are not available to the user programmer and are left to the operating system to use.

LSL	Load segment limit
LTR	Load task register
SGDT	Store global descriptor table
SIDT	Store interrupt descriptor table
STR	Store task register
SLDT	Store local descriptor table
SMSW	Store machine status word
VERR	Verify segment for reading
VERW	Verify segment for writing

Addressing modes provided are:

Register direct	(Register contains operand)
Immediate	(Instruction contains data)
Displacement	(8/16 bits)
Base address	(Uses BX or BP register)
Index	(Uses DI or SI register)

80387 floating point coprocessor

The 80386 can also be used with the 80387 floating point coprocessor to provide acceleration for floating point calculations. If the device is not present, it is possible to emulate the floating point operations in software, but at a far lower performance.

Feature comparison

There is a derivative of the 80386DX called the 80386SX which provides a lower cost device while retaining the same architecture. To reduce the cost, it uses an external 16 bit data bus and a 24 bit memory bus. The SX device is not pin compatible with the DX device.

In addition, Intel have produced an 80386SL device for portable PCs which incorporates a power control module that provides support for efficient power conservation.

Although Intel designed the 80386 series, the processor has been successfully cloned by other manufacturers (both technically and legally) such as AMD, SGS Thomson, and Cyrix. Their versions are available at far higher clock speeds than the Intel originals and many PCs are now using them.

Feature	i386SX	i386DX	i386SL
Address bus	24 bit	32 bit	24 bit
Data bus	16 bit	32 bit	16 bit
FPU present	No	No	No
Memory management	Yes	Yes	Yes
Cache on-chip	No	No	Control
Branch acceleration	No	No	No
TLB support	No	No	No
Superscalar	No	No	No
Frequency (MHz)	16,20,25,33	16,20,25,33	16,20,25
Avg. cycles/inst.	4.9	4.9	<4.9
Frequency of FPU	=CPU	=CPU	=CPU
Address range	16 Mbytes	4 Gbytes	16 Mbytes
Freq. scalability	No	No	No
Transistors	275000	275000	855000
Voltage	5 v	5 v	3 v or 5 v
System management	No	No	Yes

INTEL 80486

The Intel 80486 processor is essentially an enhanced 80386. It has a similar instruction set and register model but to dismiss it as simply a go-faster 80386 would be ignoring the other features that it uses to improve performance.

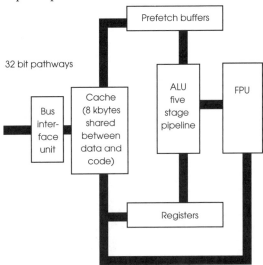

The 80486 internal architecture

Like the MC68040, it is a CISC processor that can execute instructions in a single cycle. This is done by pipelining the instruction flow so that address calculations and so on are performed as the instruction proceeds down the line. Although the pipeline may take several cycles, an instruction can potentially be started and completed on every clock edge, thus achieving the single cycle performance.

To provide instruction and data to the pipeline, the 80486 has an internal unified cache to contain both data and instructions. This removes the dependency of the processor on faster external memory to maintain sufficient data flow to allow the processor to continue executing instead of stalling. The 80486 also integrates a 80387 compatible fast floating point unit and thus does not need an external coprocessor.

Instruction set

The instruction set is essentially the same as the 80386 but there are some additional instructions available when running in protected mode to control the memory management and floating point units.

Intel 486SX and overdrive processors

The 80486 is available in several different versions which offer different facilities. The 486SX is like the 80386SX, a stripped down version of the full DX processor with the floating point unit removed but with the normal 32 bit external data and address buses. The DX2 versions are the clock doubled versions which run the internal processor at twice the external bus speed. This allows a 50 MHz DX2 processor to work in a 25 MHz board design, and opens the way to retrospective upgrades — known as the overdrive philosophy — where a user simply replaces a 25 MHz 486SX with a DX to get floating point support or a DX2 to get the FPU and theoretically twice the performance. Such upgrades need to be carefully considered: removing devices that do not have a zero insertion force socket can be tricky at best and wreck the board at worst. Similarly, the additional heat and power dissipation has also to be taken into consideration. While some early PC designs had difficulties in these areas, the overdrive option has now become a standard PC option.

The DX2 typically gives about 1.6 to 1.8 performance improvement depending on the operations that are being carried out. Internal processing gains the most from the DX2 approach while memory-intensive operations are frequently limited by the external board design.

Intel have also released a DX4 version which offers internal CPU speeds of 75 and 100 Mhz.

Feature	i486DX2-40	i486DX2-50	i486DX2-66
Address bus	32 bit	32 bit	32 bit
Data bus	32 bit	32 bit	32 bit
FPU present	Yes	Yes	Yes
Memory management	Yes	Yes	Yes
Cache on-chip	8Kunified	8K unified	8K unified
Branch acceleration	No	No	No
TLB support	No	No	No
Superscalar	No	No	No
Frequency (MHz)	40	50	66
Avg. cycles/inst.	1.03	1.03	1.03

Frequency of FPU	CPU	CPU	CPU
Upgradable	Yes	Yes	Yes
Address range	4 Gbytes	4 Gbytes	4 Gbytes
Freq. scalability	No	No	No
Transistors	1.2 million	1.2 million	1.2 million
Voltage	5 V and 3 V	5 V and 3 V	5 V

Feature	i486SX	i486DX	i486DX-50
Address bus	32 bit	32 bit	32 bit
Data bus	32 bit	32 bit	32 bit
FPU present	No	Yes	Yes
Memory management	Yes	Yes	Yes
Cache on-chip	8K unified	8K unified	8K unified
Branch acceleration	No	No	No
TLB support	No	No	No
Superscalar	No	No	No
Frequency (MHz)	16, 20, 25, 33	25, 33	50
Avg. cycles/inst.	1.03	1.03	1.03
Frequency of FPU	N/A	=CPU	=CPU
Upgradable	Yes	Yes	NO
Address range	4 Gbytes	4 Gbytes	4 Gbytes
Freq. scalability	NO	NO	NO
Transistors	1.2 million	1.2 million	1.2 million
Voltage	5 V and 3 V	5 V and 3 V	5 V
System management	No	No	No

Intel Pentium

The Pentium is essentially an enhanced 80486 from a programming model. It uses virtually the same programming model and instruction set — although there are some new additions.

The most noticeable enhancement is its ability to operate as a superscalar processor and execute two instructions per clock. To do this it has incorporated many new features that were not present on the 80486.

As the internal architecture diagram shows, the device has two five-stage pipelines that allow the joint execution of two integer instructions provided that they are simple enough not to use microcode or have data dependencies. This restriction is not that great a problem as many compilers have now started to concentrate on the simpler instructions within CISN instruction sets to improve their performance.

To maintain the throughput, the unified cache that appeared on the 80486 has been dropped in favour of two separate 8 kbyte caches: one for data and one for code. These caches are fed by an external 64 bit wide burst mode type data bus. The caches also now support writeback MESI policies instead of the less efficient write-through design.

Branches are accelerated using a branch target cache and work in conjunction with the code cache and prefetch buffers. The instruction set now supports an 8 byte compare and exchange instruction and a special processor identification instruction. The cache coherency support also has some new instructions to allow programmer's control of the MESI coherency policy.

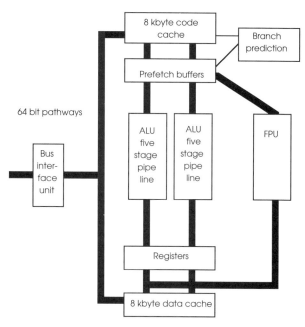

The Intel Pentium internal architecture

The Pentium has an additional control register and the system management mode register that first appeared on the 80386SL which provides intelligent power control.

Feature	Pentium
Address bus	32 bit
Data bus	64 bit
FPU present	Yes
Memory management	Yes
Cache on-chip	Two 8 K caches (data and code)
Branch acceleration	Yes — branch target cache
Cache coherency	MESI protocol
TLB support	Yes
Superscalar	Yes (2)
Frequency (MHz)	60, 66, 75, 100
Avg. cycle/inst.	0.5
Frequency of FPU	=CPU
Address range	4 Gbytes
Frequency scalability	No
Transistors	3.21 million
Voltage	5 V and 3 V
System management	Yes

Pentium Pro

The Pentium Pro processor is the name for Intel's successor to the Pentium and was previously referred to as the P6 processor. The processor was Intel's answer to the RISC threat posed by the DEC Alpha and Motorola PowerPC chips. It embraces many techniques that are used to achieve superscalar performance that have appeared previously on RISC processors such as the MPC604 PowerPC chip. It was unique in that the device actually consisted of two separate die within a single ceramic pin grid array: one die is the processor with its level one cache on-chip and the second die is the level 2 cache which is needed to maintain the instruction and data throughput needed to maintain the level of performance. It was originally introduced at 133 MHz and gained some acceptance within high-end PC/workstation applications. Both caches support the full MESI cache coherency protocol.

It achieves superscalar operation by being able to execute up to five instructions concurrently although this is a peak figure and a more realistic one to two instructions per clock is a more accurate average figure. This is done through a technique called Dynamic Execution using multiple branch prediction, dataflow analysis and speculative execution.

Multiple branch prediction

This is where the processor predicts where a branch instruction is likely to change the program flow and continues execution based on this assumption, until proven or more accurately, until correctly evaluated. This removes any delay providing the branch prediction was correct and speeds up branch execution.

Data flow anaysis

This is out of order execution and involves analysing the code and determining if there are any data dependencies. If there are, this will stall the superscalar execution until these are resolved. For example, the processor cannot execute two consecutive integer instructions if the second instruction depends on the result of its predecessor. By internally reordering the instructions, such delays can be removed and thus faster execution can be restored.

Speculative execution

Speculative execution is where instructions are executed speculatively, usually following predicted branch paths in the code until the true and correct path can be determined. If the processor has speculated correctly, then performance has been gained. If not, the speculative results are discarded and the processor continues down the correct path. If more correct speculation is achived than incorrect, the processor performance increases.

It is fair to say that the processor has not been as successful as the Pentium. This is because faster Pentium designs, especially those from Cyrix, outperformed it and were considerably cheaper. The final problem it had was that it was not optimised for 16 bit software such as MSDOS and Windows 3.x applications and required 32 bit software to really achieve its performance. The delay in the market in getting 32 bit software — Windows 95 was almost 18 months late and this stalled the market considerably — did not help its cause, and the part is now overshadowed by the faster Pentium parts and the Pentium II.

The MMX instructions

The MMX instructions or multimedia extensions as they have also been referred to were introduced to the Pentium processor to provide better support for multimedia software running on a PC using the Intel architecture. Despite some overexaggerted claims of 400% improvement, the additional instructions do provide additional support for programmers to improve their code. About 50 instructions have been added that use the SIMD (single instruction, multiple data) concept to process several pieces of data with a single instruction, instead of the normal single piece of data.

To do this, the eight floating point registers can be used to support MMX instructions or floating point. These registers are 80 bits wide and in their MMX mode, only 64 bits are used. This is enough to store a new data type known as the packed operand. This supports eight packed bytes, four packed 16 bit words, two packed 32 bit doublewords, or a single 64 bit quadword. This is extremely useful for data manipulation where pixels can be packed into the floating point register and manipulated simultaneously.

The beauty of this technique is that the basic architecture and register set does not change. The floating point registers will be saved on a context switch anyway, irrespective of whether they are storing MMX packed data or traditional floating point values. This is where one of the problems lies. A program can really only use floating point or MMX instructions. It cannot mix them without clearing the registers or saving the contents. This is because the floating point and MMX instructions share the same registers.

This has led to problems with some software and the discovery of some bugs in the silicon (run a multimedia application and then watch Excel get all the financial calculations wrong). There are fixes available and this problem will be resolved. However, the success of MMX does seem to be dependent on factors other than the technology and the MMX suffix has become a requirement. If a PC doesn't have MMX, it is no good for multimedia.

What is interesting is that MMX processors also have other improvements to help the general processor performance and so it can be a little difficult to see how much MMX can actually help. The second point is that many RISC processors, especially the PowerPC as used in the Apple Macintosh, can beat an MMX processor running the same multimedia application. The reason is simple. Many of the instructions and data manipulation that MMX brings, these processors have had as standard. They may not have packed data, but they don't have to remember if they used a floating point instruction recently and should they save the registers before using an MMX instruction. What seems to be an elegant solution does have some drawbacks.

The Pentium II

The Pentium II is the latest Intel processor and uses a module based technology and a PCB connector to provide the connection to a Intel designed motherboard. It no longer uses a chip package and is only available as a module. Essentially, a redesigned and improved Pentium Pro core with larger caches, it is the fastest Intel processor available. It is clear that Intel is focusing the PC market with its 80x86 architecture and this does raise the question the suitability of these processors to be used in embedded systems. Only time will tell if this is the case.

Motorola MC68000

The MC68000 was a complete design from scratch with the emphasis on providing an architecture that looked forward without the restrictions of remaining compatible with past designs.

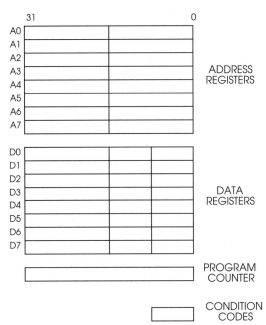

The MC68000 USER programmer's model

The only support for the old MC6800 family was a hardware interface to allow the new processor to use the existing M6800 peripherals while new M68000 parts were being designed.

Its design took many of the then current mini and mainframe computer architectural concepts and developed them using VLSI silicon technology. The programmer's register model shows how dramatic the change was. Gone are the dedicated 8 and 16 bit registers to be replaced by two groups of eight data registers and eight address registers. All these registers and the program counter are 32 bits wide.

The MC68000 hardware

Address bus

The address bus, signals A1 – A23, is non-multiplexed and 24 bits wide, giving a single linear addressing space of 16 Mbytes. A0 is not brought out directly but is internally decoded to generate upper and lower data strobes. This allows the processor to access either or both the upper and lower bytes that comprise the 16 bit data bus.

Data bus

The data bus, D0 – D15, is also non-multiplexed and provides a 16 bit wide data path to external memory and peripherals. The processor can use data in either byte, word (16 bit) or long word (32 bit) values. Both word and long word data is stored on the appropriate boundary, while bytes can be stored anywhere. The diagram shows how these data quantities are stored. All addresses specify the byte at the start of the quantity.

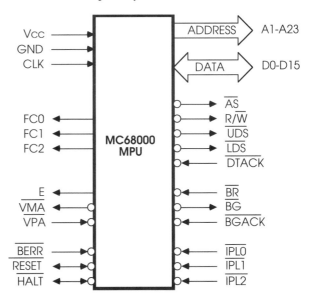

The MC68000 pinout

If an instruction needs 32 bits of data to be accessed in external memory, this is performed as two successive 16 bit accesses automatically. Instructions and operands are always 16 bits in size and accessed on word boundaries. Attempts to access instructions, operands, words or long words on odd byte boundaries cause an internal 'address' error.

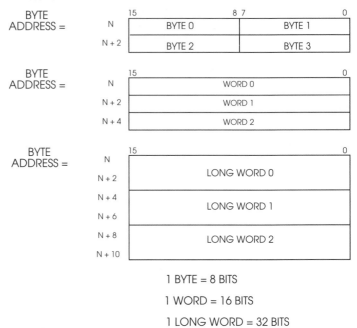

1 BYTE = 8 BITS

1 WORD = 16 BITS

1 LONG WORD = 32 BITS

MC68000 data organisation

Function codes

The function codes, FC0–FC2, provide extra information describing what type of bus cycle is occurring. These codes and their meanings are shown in the table. They appear at the same time as the address bus data and indicate program/data and supervisor/user accesses. In addition, when all three signals are asserted, the present cycle is an interrupt acknowledgement, where an interrupt vector is passed to the processor. Many designers use these codes to provide hardware partitioning.

Interrupts

Seven interrupt levels are supported and are encoded on to three interrupt pins IP0–IP2. With all three signals high, no external interrupt is requested. With all three asserted, a non-maskable level 7 interrupt is generated. Levels 1–6, generated by other combinations, can be internally masked by writing to the appropriate bits within the status register.

The interrupt cycle is started by a peripheral generating an interrupt. This is usually encoded using a 148 priority encoder. The appropriate code sequence is generated and drives the interrupt pins. The processor samples the levels and requires the levels to remain constant to be recognised. It is recommended that the interrupt level remains asserted until its interrupt acknowledgement cycle commences to ensure recognition. Once the processor has recognised the interrupt, it waits until the current instruction has been completed and starts an interrupt acknowledgement cycle. This starts an external bus cycle with all three function codes driven high to indicate an interrupt acknowledgement cycle.

Function code			Reference class
FC0	FC1	FC2	
0	0	0	Reserved
0	0	1	User data
0	1	0	User program
0	1	1	Reserved (I/O space)
1	0	0	Reserved
1	0	1	Supervisor data
1	1	0	Supervisor program
1	1	1	CPU space/ interrupt ack

The MC68000 function codes and their meanings

The interrupt level being acknowledged is placed on address bus bits A1–A3 to allow external circuitry to identify which level is being acknowledged. This is essential when one or more interrupt requests are pending. The system now has a choice over which way it will respond:

- If the peripheral can generate an 8 bit vector number, this is placed on the lower byte of the address bus and DTACK* asserted. The vector number is read and the cycle completed. This vector number then selects the address and subsequent software handler from the vector table.

- If the peripheral cannot generate a vector, it can assert VPA* and the processor will terminate the cycle using the M6800 interface. It will select the specific interrupt vector allocated to the specific interrupt level. This method is called auto-vectoring.

To prevent an interrupt request generating multiple acknowledgements, the internal interrupt mask is raised to the interrupt level, effectively masking any further requests. Only if a higher level interrupt occurs will the processor nest its interrupt service routines. The interrupt service routine must clear the interrupt source and thus remove the request before returning to normal execution. If another interrupt is pending from a different source, it will be recognised and cause another acknowledgement to occur.

Error recovery and control signals

There are three signals associated with error control and recovery. The bus error BERR*, HALT* and RESET* signals can provide information or be used as inputs to start recovery procedures in case of system problems.

The BERR* signal is the counterpart of DTACK*. It is used during a bus cycle to indicate an error condition that may arise through parity errors or accessing non-existent memory. If BERR* is asserted on its own, the processor halts normal processing and goes to a special bus error software handler. If HALT* is asserted at the same time, it is possible to rerun the bus cycle. BERR* is removed followed by HALT* one clock later, after which the previous cycle is rerun automatically. This is useful to screen out transient errors. Many designs use external hardware to force a rerun automatically but will cause a full bus error if an error occurs during the rerun.

Without such a signal, the only recourse is to complete the transfer, generate an immediate non-maskable interrupt and let a software handler attempt to sort out the mess! Often the only way out is to reset the system or shut it down. This makes the system extremely intolerant of signal noise and other such transient errors.

The RESET* and HALT* signals are driven low at power-up to force the MC68000 into its power-up sequence. The operation takes about 100 ms, after which the signals are negated and the processor accesses the Reset vector at location 0 in memory to fetch its stack pointer and program counter from the two long words stored there.

Motorola MC68020

The MC68020 was launched in April 1984 as the '32 bit performance standard' and in those days its performance was simply staggering — 8 million instructions per second peak with 2–3 million sustained when running at 16 MHz clock speed. It was a true 32 bit processor with 32 bit wide external data and address buses as shown. It supported all the features and functions of the MC68000 and MC68010, and it executed M68000 USER binary code without modification (but faster!).

- Virtual memory and instruction continuation were supported. This is explained in Chapter 7 on interrupts.

- The bus and control signals were similar to that of its M68000 predecessors, offering an asynchronous memory interface but with a three–cycle operation (instead of four) and dynamic bus sizing which allowed the processor to talk to 8, 16 and 32 bit processors.

- Additional coprocessors could be added to provide such facilities as floating point arithmetic and memory management, which used this bus to provide a sophisticated communications interface.

- The instruction set was enhanced with more data types, addressing modes and instructions.
- Bit field data and its manipulation was supported, along with packed and unpacked BCD (binary coded decimal) formats. An instruction cache and a barrel shifter to perform high speed shift operations were incorporated on-chip to provide support for these functions.

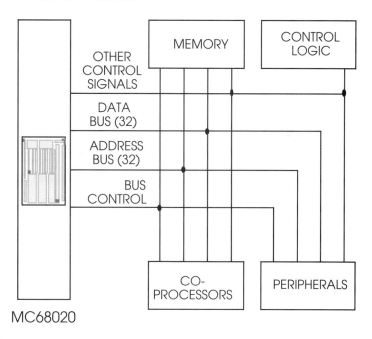

MC68020

A simple MC68020 system

The actual pipeline used within the design is quite sophisticated. It is a four–stage pipe with stage A consisting of an instruction router which accepts data from either the external bus controller or the internal cache. As the instruction is processed down the pipeline, the intermediate data can either cause micro and nanocode sequences to be generated to control the execution unit or, in the case of simpler instructions, the data itself can be passed directly into the execution unit with the subsequent speed improvements.

The programmer's model

The programmer's USER model is exactly the same as for the MC68000, MC68010 and MC68008. It has the same eight data and eight address 32 bit register organisation. The SUPERVISOR mode is a superset of its predecessors. It has all the registers found in its predecessors plus another three. Two registers are associated with controlling the instruction cache, while the third provides the master stack pointer.

The supervisor uses either its master stack pointer or interrupt stack pointer, depending on the exception cause and the status of the M bit in the status register. If this bit is clear, all stack operations default to the A7´ stack pointer. If it is set, interrupt stack frames are stored using the interrupt stack pointer while other operations use the master pointer. This effectively allows the system to maintain two separate stacks. While primarily for operating system support, this extra register can be used for high reliability designs.

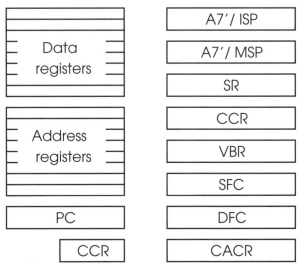

The MC68040 programming model

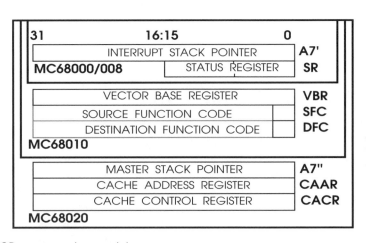

The M68020 SUPERVISOR programming model

The MC68020 instruction set is a superset of the MC68000/ MC68010 sets. The main difference is the inclusion of floating point and coprocessor instructions, together with a set to manipulate bit field data. The instructions to perform a trap on condition operation, a compare and swap operation and a 'call-return from module' structure were also included. Other differences were the addition of 32 bit displacements for the LINK and Bcc (branch on condition) instructions, full 32 bit arithmetic with 32 or 64 bit results as appropriate and extended bounds checking for the CHK (check) and CMP (compare) instructions.

The bit field instructions were included to provide additional support for applications where data does not conveniently fall into a byte organisation. Telecommunications and graphics both manipulate data in odd sizes — serial data can often be 5, 6 or 7 bits in size and graphics pixels (i.e. each individual dot that makes a picture on a display) vary in size, depending on how many colours, grey scales or attributes are being depicted.

Dn	Data Register Direct
An	Address Register Direct
(An)+	Address Reg. Indirect w/ Post-Increment
-(An)	Address Reg. Indirect w/ Pre-Decrement
d(An)	Displaced Address Register Indirect
d(An,Rx)	Indexed, Displaced Address Reg.
d(PC)	Program Counter Relative
d(PC,Rx)	Indexed Program Counter Relative
#xxxxxxxx	Immediate
$xxxx	Absolute Short
$xxxxxxxx	Absolute Long

MC68000/008/010

(bd,An,Xn.SIZE*SCALE)	Register Indirect
	Memory Indirect
((bd,An,Xn.SIZE*SCALE),od)	Pre-Indexed
((bd,An),Xn.SIZE*SCALE,od)	Post-Indexed
	Program Counter Memory Indirect
((bd,PC,Xn.SIZE*SCALE),od)	Pre-Indexed
((bd,PC),Xn.SIZE*SCALE,od)	Post-Indexed

MC68020/MC68030

The MC68020 addressing modes

The addressing modes were extended from the basic M68000 modes, with memory indirection and scaling. Memory indirection allowed the contents of a memory location to be used within an effective address calculation rather than its absolute address. The scaling was a simple multiplier value 1, 2, 4 or 8 in magnitude, which multiplied (scaled) an index register. This allowed large data elements within data structures to be easily accessed without having to perform the scaling calculations prior to the access. These new modes were so complex that even the differentiation between data and address registers was greatly reduced: with the MC68020, it is possible to use data registers as additional address registers. In practice, there are over 50 variations available to the programmer to apply to the 16 registers.

The new CAS and CAS2 'compare and swap' instructions provided an elegant solution to linked list updating within a multiprocessor system. A linked list is a series of data lists linked together by storing the address of the next list in the chain in the preceding chain. To add or delete a list simply involves modifying these addresses.

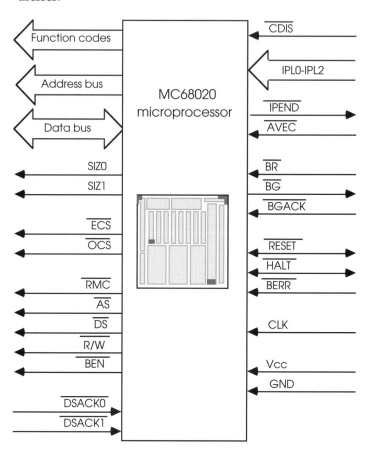

The MC68020 device pinout

In a multiprocessor system, this modification procedure must occur uninterrupted to prevent corruption. The CAS and CAS2 instruction meets this specification. The current pointer to the next list is read and stored in Dn. The new value is calculated and stored in Dm. The CAS instruction is then executed. The current pointer value is read and compared with Dn. If they are the same, no other updating by another processor has happened and Dm is written out to update the list. If they do not match, the value is copied into Dn, ready for a repeat run. This sequence is performed using an indivisible read-modify-write cycle. The condition codes are updated during each stage. The CAS2 instruction performs a similar function but with two sets of values. This instruction is also performed as a series of indivisible cycles but with different addresses appearing during the execution.

Bus interfaces

Many of the signals shown in the pin out diagram are the same as those of the MC68000 — the function codes FC0–2, interrupt pins IPL0–2 and the bus request pins, RESET*, HALT* and BERR* perform the same functions.

With the disappearance of the M6800 style interface, separate signals are used to indicate an auto-vectored interrupt. The AVEC* signal is used for this function and can be permanently asserted if only auto-vectored interrupts are required. The IPEND signal indicates when an interrupt has been internally recognised and awaits an acknowledgement cycle. RMC* indicates an indivisible read-modify-write cycle instead of simply leaving AS* asserted between the bus cycles. The address strobe is always released at the end of a cycle. ECS* and OCS* provide an early warning of an impending bus cycle, and indicate when valid address information is present on the bus prior to validation by the address strobe.

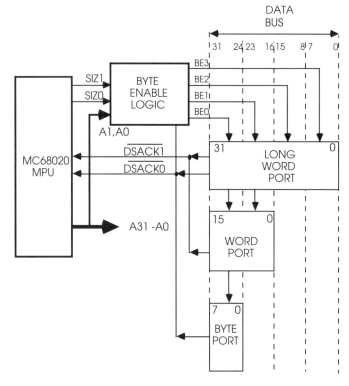

DSACK1	DSACK0	MEANING
HI	HI	Insert wait state
HI	LO	Complete cycle, port size = 8 bits
LO	HI	Complete cycle, port size = 16 bits
LO	LO	Complete cycle, port size = 32 bits

MC68020 dynamic bus sizing

The M68000 upper and lower data strobes have been replaced by A0 and the two size pins, SIZE0 and SIZE1. These indicate the amount of data left to transfer on the current bus cycle and, when used with address bits A0 and A1, can provide decode information so that the correct bytes within the 4 byte wide data bus can be enabled. The old DTACK* signal has been replaced by two new ones, DSACK0* and DSACK1*. They provide the old DTACK* function of indicating a successful bus cycle and are used in the dynamic bus sizing. The bus interface is asynchronous and similar to the M68000 cycle but with a shorter three–cycle sequence, as shown.

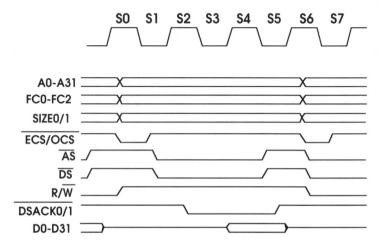

MC68020 bus cycle timings

Motorola MC68030

The MC68030 appeared some 2–3 years after the MC68020 and used the advantage of increased silicon real estate to integrate more functions on to a MC68020-based design. The differences between the MC68020 and the MC68030 are not radical — the newer design can be referred to as evolutionary rather than a quantum leap. The device is fully MC68020 compatible with its full instruction set, addressing modes and 32 bit wide register set. The initial clock frequency was designed to 20 MHz, some 4 MHz faster than the MC68020, and this has yielded commercially available parts running at 50 MHz. The transistor count has increased to about 300000 but with its smaller geometries, die size and heat dissipation are similar.

Memory management has now been brought on-chip with the MC68030 using a subset of the MC68851 PMMU with a smaller 22 entry on-chip address translation cache. The 256 byte instruction cache of the MC68020 is still present and has been augmented with a 256 byte data cache.

Both these caches are logical and are organised differently from the 64 × 4 MC68020 scheme. A 16 × 16 organisation has been adopted to allow a new synchronous bus to burst fill cache lines. The cache lookup and address translations occur in parallel to improve performance.

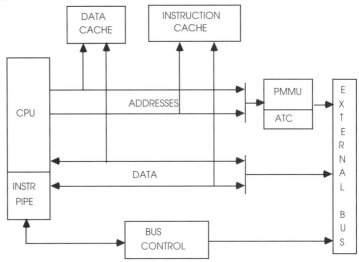

The MC68030 internal block diagram

The processor supports both the coprocessor interface and the MC68020 asynchronous bus with its dynamic bus sizing and misalignment support. However, it has an alternative synchronous bus interface which supports a two–clock access with optional single-cycle bursting. The bus interface choice can be made dynamically by external hardware.

The MC68040

The MC68040 incorporates separate integer and floating point units giving sustained performances of 20 integer MIPS and 3.5 double precision Linpack MFLOPS respectively, dual 4 kbyte instruction and data caches, dual memory management units and an extremely sophisticated bus interface unit. The block diagram shows how the processor is partitioned into several separate functional units which can all execute concurrently. It features a full Harvard architecture internally and is remarkably similar at the block level, to the PowerPC RISC processor.

The design is revolutionary rather than evolutionary: it takes the ideas of overlapping instruction execution and pipelining to a new level for CISC processors. The Floating Point and integer execution units work in parallel with the on-chip caches and memory management to increase the overlapping so that many instructions are executed in a single cycle, and thus give it its performance.

The pinout reveals a large number of new signals. One major difference about the MC68040 is its high drive capability. The processor can be configured on reset to drive either 55 or 5 mA per bus or control pin. This removes the need for externals buffers, reducing chip count and the associated propagation delays, which often inflict a high speed design. The 32 bit address and 32 bit data buses are similar to its predecessors although the signals can be optionally tied together to form a single 32 bit multiplexed data/address bus.

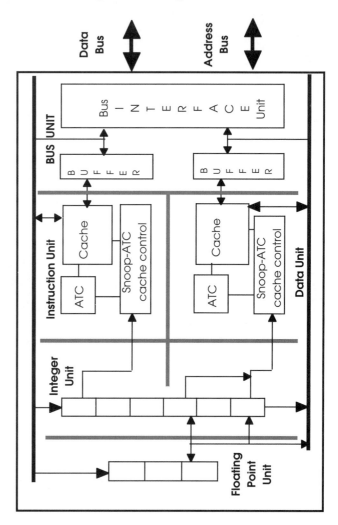

The MC68040 block diagram

The User Programmable Attributes (UPA0 and UPA1) are driven according to 2 bits within each page descriptor used by the onboard memory management units. They are primarily used to enable the MC68040 Bus Snooping protocols, but can also be used to give additional address bits, software control for external caches and other such functions. The two size pins, SIZ0 and SIZ1, no longer indicate the number of remaining bytes left to be transferred as they did on the MC68020 and MC68030, but are used to generate byte enables for memory ports. They now indicate the size of the current transfer. Dynamic bus sizing is supported via external hardware if required. Misaligned accesses are supported by splitting the transfer into a series of aligned accesses of differing sizes. The transfer type signals, TT1 and TT2, indicate the type of transfer that is taken place and the Transfer Modifier pins TM0-2 provide further information. These five pins effectively replace the three function code pins. The TLN0-1 pins indicate the current long word number within a burst fill access.

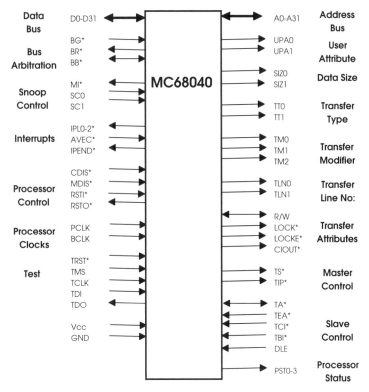

The MC68040 pinout

The synchronous bus is controlled by the Master and Slave transfer control signals: Transfer Start TS* indicates a valid address on the bus while the Transfer in Progress TIP* signal is asserted during all external bus cycles and can be used to power up/down external memory to conserve power in portable applications. These two Master signals are complemented by the slave signals: Transfer Acknowledge TA* successfully terminates the bus cycle, while Transfer Error Acknowledge TEA* terminates the cycle and the burst fill as a result of an error. If both these signals are asserted on the first access of the burst, the cycle is terminated and immediately rerun. On the 2nd, 3rd and 4th accesses, a retry attempt is not allowed and the processor simply assumes that an error has occurred and will terminate the burst as normal.

The processor can be configured to use a different signal, Data Latch Enable DLE to latch read data instead of the rising edge of the BCLK clock. The internal caches and memory management units can be disabled via the CDIS* and MDIS* pins respectively.

The programming model

To the programmer the programming model of the MC68040 is the same as its predecessors such as the MC68030. It has the same 8 data and 8 address registers, the vector base register VBR, the alternate function code registers although some codes are reserved, the dual Supervisor stack pointer and the two cache control registers although only two bits are now used to enable or disable either of the two on-chip caches. Internally the implementation is different. Its instruction execution unit consists of a six–stage pipeline which sequentially fetches an instruction, decodes it, calculates the effective address, fetches an address operand, executes the instruction and finally writes back the results. To prevent pipeline stalling, an internal Harvard architecture is used to allow simultaneous instruction and operand fetches. It has been optimised for many instructions and addressing modes so that single-cycle execution can be achieved. The early pipeline stages are effectively duplicated to allow both paths of a branch instruction to be processed until the path decision is taken. This removes pipeline stalls and the subsequent performance degradation. While integer instructions are being executed, the floating point unit is free to execute floating point instructions.

Data registers		A7' / ISP
		A7' / MSP
		SR
Address registers		CCR
		VBR
		SFC
PC		DFC
CCR		CACR

The MC68040 programming model

Integrated processors

With the ability of semiconductor manufacturers to be able to integrate several million transistors onto a single piece of silicon, it should come as no surprise that there are now processors available which take the idea of integration offered by a microcontroller, but use a high performance processor instead. The Intel 80186 started this process by combining DMA channels with a 8086 architecture. The most successful family so far has been the MC683xx family from Motorola. There are now several members of the family currently available.

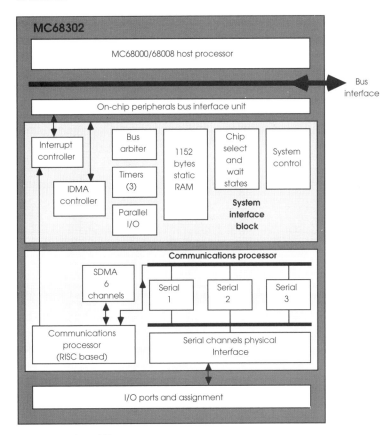

The MC68302 Integrated Multiprotocol Processor

They combine a M68000 or MC68020 (CPU32) family processor and its asynchronous memory interface, with all the standard interface logic of chip selects, wait state generators, bus and watchdog timers, into a system interface module and use a second RISC type processor to handle specific I/O functions. This approach means that all the additional peripherals and logic needed to construct a MC68000-based system has gone. In many cases, the hardware design is almost at the 'join up the dots' level where the dots are the processor and memory pins.

This approach has been adopted by others and many different processor cores, such as SPARC and MIPs, are now available in similar integrated processors. PowerPC versions of the MC68360 are now in production where the MC68000-based CPU32 core is replaced with a 50 MHz PowerPC processor. For embedded systems, this is definitely the way of the future.

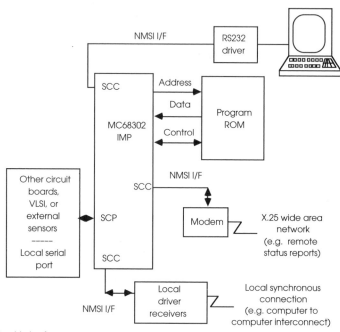

A typical X25-ISDN terminal interface

The MC68302 uses a 16 MHz MC68000 processor core with power down modes and either an 8 or 16 bit external bus. The system interface block contains 1152 bytes of dual port RAM, 28 pins of parallel I/O, an interrupt controller and a DMA device, as well as the standard glue logic. The communications processor is a RISC machine that controls three multiprotocol channels, each with their own pair of DMA channels. The channels support BISYNC, DDCMP, V.110, HDLC synchronous modes and standard UART functions. This processor takes buffer structures from either the internal or external RAM and takes care of the day-to-day activities of the serial channels. It programs the DMA channel to transfer the data, performs the character and address comparisons and cyclic redundancy check (CRC) generation and checking. The processor has sufficient power to cope with a combined data rate of 2 Mbits per second across the three channels. Assuming an 8 bit character and a single interrupt to poll all three channels, the processor is handling the equivalent of an interrupt every 12 microseconds. In addition, it is performing all the data framing etc. While this is going on, the on-chip M68000 is free to perform other functions, like the higher layers of X.25 or other OSI protocols as shown.

The MC68332 is similar to the MC68302, except that it has a CPU32 processor (MC68020 based) running at 16 MHz and a timer processor unit instead of a communications processor. This has 16 channels which are controlled by a RISC-like processor to perform virtually any timing function. The timing resolution is down to 250 nanoseconds with an external clock source or 500 nanoseconds with an internal one. The timer processor can perform the common timer algorithms on any of the 16 channels without placing any overhead on the CPU32.

A queued serial channel and 2 kbits of power down static RAM are also on-chip and for many applications, all that is required to complete a working system is an external program EPROM and a clock.

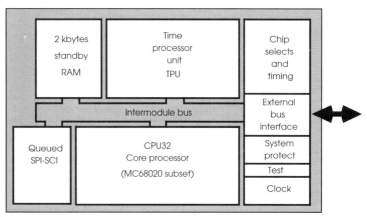

The MC68332 block diagram

This is a trend that many other architectures are following especially with RISC processors. Apart from the high performance range of the processor market or where complete flexibility is needed, most processors today come with at least some basic peripherals such as serial and parallel ports and a simple or glueless interface to memory. In many cases, they dramatically reduce the amount of hardware design needed to add external memory and thus complete a simple design. This type of processor is gaining popularity with designers.

RISC processors

Until 1986, the expected answer to the question 'which processor offers the most performance' would be MC68020, MC68030 or even 386! Without exception, CISC processors such as these, had established the highest perceived performances. There were more esoteric processors, like the transputer, which offered large MIPS figures from parallel arrays but these were often considered only suitable for niche markets and applications. However, around this time, there started an interest in an alternative approach to micro-processor design, which seemed to offer more processing power from a simpler design using less transistors. Performance increases of over five times the then current CISC machines were suggested. These machines, such as the Sun SPARC architecture and the MIPS R2000 processor, were the first of a modern generation of processors based on a reduced instruction set, generically called reduced instruction set processors (RISC).

The 80/20 rule

Analysis of the instruction mix generated by CISC compilers is extremely revealing. Such studies for CISC mainframes and mini computers shows that about 80% of the instructions generated and executed used only 20% of an instruction set. It was an obvious conclusion that if this 20% of instructions was speeded up, the performance benefits would be far greater. Further analysis shows that these instructions tend to perform the simpler operations and use only the simpler addressing modes. Essentially, all the effort invested in processor design to provide complex instructions and thereby reduce the compiler workload was being wasted. Instead of using them, their operation was synthesised from sequences of simpler instructions.

This has another implication. If only the simpler instructions are required, the processor hardware required to implement them could be reduced in complexity. It therefore follows that it should be possible to design a more performant processor with fewer transistors and less cost. With a simpler instruction set, it should be possible for a processor to execute its instructions in a single clock cycle and synthesise complex operations from sequences of instructions. If the number of instructions in a sequence, and therefore the number of clocks to execute the resultant operation, was less than the cycle count of its CISC counterpart, higher performance could be achieved. With many CISC processors taking 10 or more clocks per instruction on average, there was plenty of scope for improvement.

The initial RISC research

The computer giant IBM is usually acknowledged as the first company to define a RISC architecture in the 1970s. This research was further developed by the Universities of Berkeley and Stanford to give the basic architectural models. RISC can be described as a philosophy with three basic tenets:

1.   All instructions will be executed in a single cycle

        This is a necessary part of the performance equation. Its imple-
        mentation calls for several features — the instruction op code
        must be of a fixed width which is equal to or smaller than the

size of the external data bus, additional operands cannot be supported and the instruction decode must be simple and orthogonal to prevent delays. If the op code is larger than the data width or additional operands must be fetched, multiple memory cycles are needed, increasing the execution time.

2.   Memory will only be accessed via load and store instructions

This naturally follows from the above. If an instruction manipulates memory directly, multiple cycles must be performed to execute it. The instruction must be fetched and memory manipulated. With a RISC processor, the memory resident data is loaded into a register, the register manipulated and, finally, its contents written out to main memory. This sequence takes a minimum of three instructions. With register-based manipulation, large numbers of general-purpose registers are needed to maintain performance.

3.   All execution units will be hardwired with no microcoding

Microcoding requires multiple cycles to load sequencers etc and therefore cannot be easily used to implement single-cycle execution units.

Two generic RISC architectures form the basis of nearly all the current commercial processors. The main differences between them concern register sets and usage. They both have a Harvard external bus architecture consisting of separate buses for instructions and data. This allows data accesses to be performed in parallel with instruction fetches and removes any instruction/data conflict. If these two streams compete for a single bus, any data fetches stall the instruction flow and prevent the processor from achieving its single cycle objective. Executing an instruction on every clock requires an instruction on every clock.

The Berkeley RISC model

The RISC 1 computer implemented in the late 1970s used a very large register set of 138×32 bit registers. These were arranged in eight overlapping windows of 24 registers each. Each window was split so that six registers could be used for parameter passing during subroutine calls. A pointer was simply changed to select the group of six registers. To perform a basic call or return simply needed a change of pointer. The large number of registers is needed to minimise the number of fetches to the outside world. With this simple window technique, procedure calls can be performed extremely quickly. This can be very beneficial for real-time applications where fast responses are necessary.

However, it is not without its disadvantages. If the procedure calls require more than six variables, one register must be used to point to an array stored in external memory. This data must be loaded prior to any processing and the register windowing loses much of its performance. If all the overlapping windows are used, the system resolves the situation by tracking the window usage so either a window or the complete register set can be saved out to external memory.

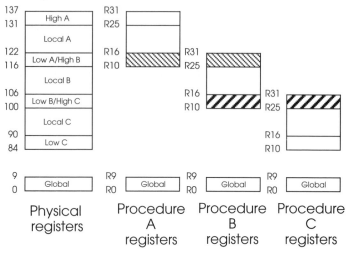

Register windowing

This overhead may negate any advantages that windowing gave in the first place. In real-time applications, the overhead of saving 138 registers to memory greatly increases the context switch and hence the response time.

A good example of this approach is the Sun SPARC processor.

Sun SPARC RISC processor

The SPARC (scalable processor architecture) processor is a 32 bit RISC architecture developed by Sun Microsystems for their workstations but manufactured by a number of manufacturers such as LSI, Cypress, Fujitsu, Philips and Texas Instruments.

The basic architecture follows the Berkeley model and uses register windowing to improve context switching and parameter passing. The initial designs were based on a discrete solution with separate floating point units, memory management facilities and cache memory, but later designs have integrated these versions. The latest versions also support superscalar operation.

Architecture

The SPARC system is based on the Berkeley RISC architecture. A large 32 bit wide register bank containing 120 registers is divided into a set of seven register windows and a set of eight registers which are globally available. Each window set containing 24 registers is split into three sections to provide eight input, eight local and eight output registers. The registers in the output section provide the information to the eight input registers in the next window. If a new window is selected during a context switch or as a procedural call, data can be passed with no overhead by placing it in the output registers of the first window. This data is then available to the procedure or next context in its input registers. In this way, the windows are linked together to form a chain where the input registers for one window have the contents of the output registers of the previous window.

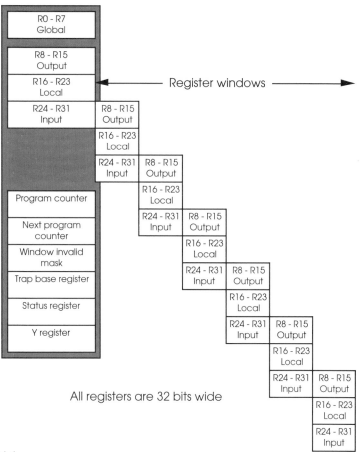

The SPARC register model

To return information back to the original or calling software, the data is put into the input registers and the return executed. This moves the current window pointer back to the previous window and the returned information is now available in that window's output registers. This method is the reverse of that used to initially pass the information in the first place.

The programmer and CPU can track and control which windows are used and what to do when all windows are full, through fields in the status register.

The architecture is also interesting in that it is one of the few RISC processors that uses logical addressed caches instead of physically addressed caches.

Interrupts

The SPARC processor supports 15 external interrupts which are generated using the four interrupt lines, IRL0 – IRL3. Level 15 is assigned as a non-maskable interrupt and the other 14 can be masked if required.

An external interrupt will generate an internal trap where the current and the next instructions are saved, the pipeline flushed and the processor switched into supervisor mode. The trap vector table which is located in the trap base register is then used to supply the address of the service routine. When the routine has completed, the REIT instruction is executed which restores the processor status and allows it to continue.

Instruction set

The instruction set comprises of 64 instructions. All access to memory is via load and store instructions as would be expected with a RISC architecture. All other instructions operate on the register set including the currently selected window. The instruction set is also interesting in that it has a multiply step command instead of the more normal multiply command. The multiply step command allows a multiply to be synthesised.

The Stanford RISC model

This model uses a smaller number of registers (typically 32) and relies on software techniques to allocate register usage during procedural calls. Instruction execution order is optimised by its compilers to provide the most efficient way of performing the software task. This allows pipelined execution units to be used within the processor design which, in turn, allow more powerful instructions to be used.

However, RISC is not the magic panacea for all performance problems within computer design. Its performance is extremely dependent on very good compiler technology to provide the correct optimisations and keep track of all the registers. Many of the early M68000 family compilers could not track all the 16 data and address registers and therefore would only use two or three. Some compilers even reduced register usage to one register and effectively based everything on stacks and queues. Secondly, the greater number of instructions it needed increased code size dramatically at a time when memory was both expensive and low in density. Without the compiler technology and cheap memory, a RISC system was not very practical and the ideas were effectively put back on the shelf.

The MPC601 was the first PowerPC processor available. It has three execution units: a branch unit to resolve branch instructions, an integer unit and a floating point unit.

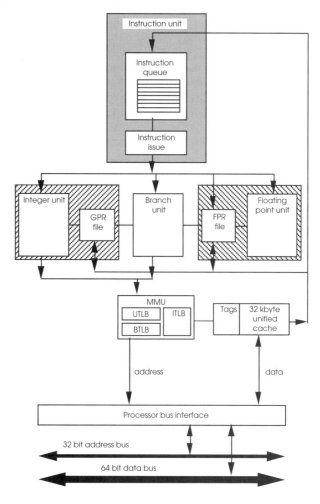

MPC601 internal block diagram

The floating point unit supports IEEE format. The processor is superscalar. It can dispatch up to two instructions and process three every clock cycle. Running at 66 MHz, this gives a peak performance of 132 million instructions per second.

The branch unit supports both branch folding and speculative execution where the processor speculates which way the program flow will go when a branch instruction is encountered and start executing down that route while the branch instruction is resolved.

The general-purpose register file consists of 32 separate registers, each 32 bits wide. The floating point register file also contains 32 registers, each 64 bits wide, to support double precision floating point. The external physical memory map is a 32 bit address linear organisation and is 4 Gbytes in size.

The MPC601's memory subsystem consists of a unified memory management unit and on-chip cache which communicates to external memory via a 32 bit address bus and a 64 bit data bus. At its peak, this bus can fetch two instructions per clock or 64 bits of data. It also supports split transactions, where the address bus can be used independently and simultaneously with the data bus to improve its utilisation. Bus snooping is also provided to ensure cache coherency with external memory.

The cache is 32 kbytes and supports both data and instruction accesses. It is accessed in parallel with any memory management translation. To speed up the translation process, the memory management unit keeps translation information in one of three translation lookaside buffers.

The MPC603 block diagram

The MPC603 was the second PowerPC processor to appear. Like the MPC601, it has the three execution units: a branch unit to resolve branch instructions, an integer unit and a floating point unit.

The floating point unit supports IEEE format. However, two additional execution units have been added to provide dedicated support for system registers and to move data between the register files and the two on-chip caches. The processor is superscalar and can dispatch up to three instructions and process five every clock cycle.

The branch unit supports both branch folding and speculative execution. It augments this with register renaming, which allows speculative execution to continue further than allowed on the MPC601 and thus increase the processing advantages of the processor.

The general-purpose register file consists of 32 separate registers, each 32 bits wide. The floating point register file contains 32 registers, each 64 bits wide to support double precision floating point. The external physical memory map is a 32 bit address linear organisation and is 4 Gbytes in size.

The MPC603's memory subsystem consists of a separate memory management unit and on-chip cache for data and instructions which communicates to external memory via a 32 bit address bus and a 64 or 32 bit data bus. This bus can, at its peak, fetch two instructions per clock or 64 bits of data. Each cache is 8 kbytes in size, giving a combined on-chip cache size of 16 kbytes. The bus also supports split transactions, where the address bus can be used independently and simultaneously with the data bus to improve its utilisation. Bus snooping is also provided to ensure cache coherency with external memory.

As with the MPC601, the MPC603 speeds up the address translation process, by keeping translation information in one of four translation lookaside buffers, each of which is divided into two pairs, one for data accesses and the other for instruction fetches. It is different from the MPC601 in that translation tablewalks are performed in software and not automatically by the processor.

The device also includes power management facilities and is eminently suitable for low power applications.

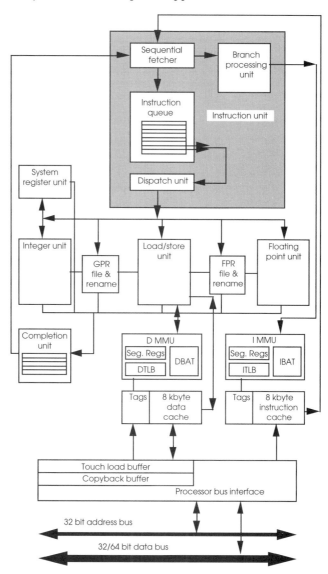

MPC603 internal block diagram

Signal processors

Signal processors started out as special processors that were designed for implementing digital signal processing (DSP) algorithms. A good example of a DSP function is the finite impulse response (FIR) filter. This involves setting up two tables, one containing sampled data and the other filter coefficients that determine the filter response. The program then performs a series of repeated multiply and accumulates using values from the tables. The bandwidth of such filters depends on the speed of these simple operations. With a general-purpose architecture like the M68000 family the code structure would involve setting up two tables in external memory, with an address register allocated to each one to act as a pointer. The beginning and the end of the code would consist of the loop initialisation and control, leaving the multiply–accumulate operations for the central part. The M68000 instruction set does offer some facilities for efficient code: the incremental addressing allows pointers to progress down the tables automatically, and the decrement and branch instruction provides a good way of implementing the loop structures. However, the disadvantages are many: the multiply takes >40 clocks, the single bus is used for all the instruction fetches and table searches, thus consuming time and bandwidth. In addition the loop control timings vary depending on whether the branch is taken or not. This can make bandwidth predictions difficult to calculate. This results in very low bandwidths and is therefore of limited use within digital signal processing. This does not mean that an MC68000 cannot perform such functions: it can, providing performance is not of an issue.

RISC architectures like the PowerPC family can offer some immediate improvements. The capability to perform single cycle arithmetic is an obvious advantage. The Harvard architecture reduces the execution time further by allowing simultaneous data and instruction fetches. The PowerPC can, by virtue of its high performance, achieve performances suitable for many DSP applications. The system cost is high involving a multiple chip solution with very fast memory etc. In applications that need high speed general processing as well, it can also be a suitable solution.

Another approach is to build a dedicated processor to perform certain algorithms. By using discrete building blocks, such as hardware multipliers, counters etc., a totally hardware solution can be designed to perform such functions. Modulo counters can be used to form the loop structures and so on. The disadvantages are cost and a loss of flexibility. Such hardware solutions are difficult to alter or program. What is obviously required is a processor whose architecture is enhanced specifically for DSP applications.

DSP basic architecture

As an example of a powerful DSP processor, consider the Motorola DSP56000. It is used in many digital audio applications where it acts as a multi-band graphics equaliser or as a noise reduction system.

The processor is split into 10 functional blocks. It is a 24 bit data word processor to give increased resolution. The device has an enhanced Harvard architecture with three separate external buses: one for program and X and Y memories for data. The communication between these and the outside world is controlled by two external bus switches, one for data and the other for addresses. Internally, these two switches are functionally reproduced by the internal data bus switch and the address arithmetic unit (AAU). The AAU contains 24 address registers in three banks of 8. These are used to reference data so that it can be easily fetched to maintain the data flow into the data ALU.

The program address generator, decode controller and interrupt controller organise the instruction flow through the processor. There are six 24 bit registers for controlling loop counts, operating mode, stack manipulation and condition codes. The program counter is 24 bit although the upper 8 bits are only used for sign extension.

The main workhorse is the data ALU, which contains two 56 bit accumulators A and B which each consist of three smaller registers A0, A1,A2, B0,B1 and B2. The 56 bit value is stored with the most significant 24 bit word in A1 or B1, the least significant 24 bit word in A0 or B0 and the 8 bit extension word is stored in A2 or B2. The processor uses a 24 bit word which can provide a dynamic range of some 140 dB, while intermediate 56 bit results can extend this to 330 dB. In practice, the extension byte is used for over- and underflow. In addition there are four 24 bit registers X1, X0, Y1 and Y0. These can also be paired to form two 48 bit registers X and Y.

These registers can read or write data from their respective data buses and are the data sources for the multiply–accumulate (MAC) operation. When the MAC instruction is executed, two 24 bit values from X0,X1,Y1 or Y0 are multiplied together, and then added or subtracted from either accumulator A or B. This takes place in a single machine cycle of 75 ns at 27 MHz. While this is executing, two parallel data moves can take place to update the X and Y registers with the next values. In reality, four separate operations are taking place concurrently.

The data ALU also contains two data shifters for bit manipulation and to provide dynamic scaling of fixed point data without modifying the original program code by simply programming the scaling mode bits. The limiters are used to reduce any arithmetic errors due to overflow, for example. If overflow occurs, i.e. the resultant value requires more bits to describe it than are available, then it is more accurate to write the maximum valid number than the overflowed value. This maximum or limited value is substituted by the data limiter in such cases, and sets a flag in the condition code register to indicate what has happened.

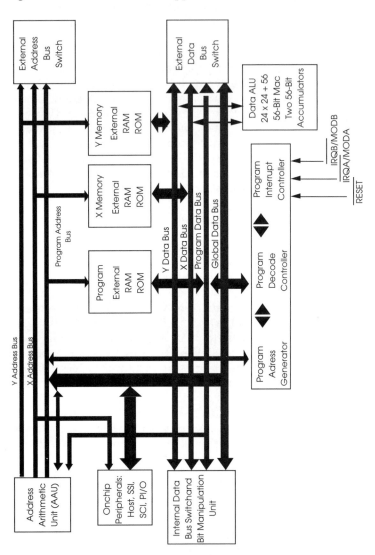

The DSP56000 block diagram

The external signals are split into various groups. There are three ports A, B and C and seven special bus control signals, two interrupt pins, reset, power and ground and, finally, clock signals. The device is very similar in design to an 8 bit microcontroller unit (MCU), and it can be set into several different memory configurations.

The three independent memory spaces, X data, Y data and program are configured by the MB,MA and DE bits in the operating mode register. The MB and MA bits are set according to the status of the MB and MA pins during the processor´s reset sequence. These pins are subsequently used for external interrupts. Within the program space, the MA and MB bits determine where the program memory is and where the reset starting address is located. The DE bit either effectively enables or disables internal data ROMs which contain a set of μ and A Law expansion tables in the X data ROM and a four quadrant sine wave table in the Y data ROM. The on-chip peripherals are mapped into the X data space between $FFC0 and $FFFF. Each of the three spaces is 64 kbytes in size.

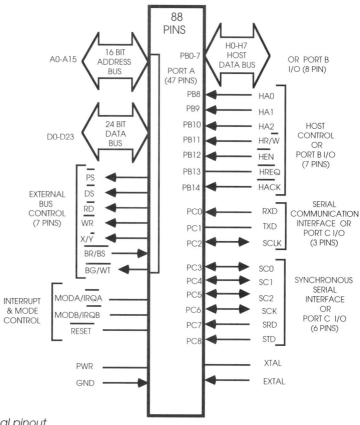

The DSP56000/1 external pinout

These memory spaces communicate to the outside world via a shared 16 bit address bus and a 24 bit data bus. Two additional signals, PS* and X/Y* identify which type of access is taking place. The DSP56000 can be programmed to insert a fixed number of wait states on external accesses for each memory space and I/O. Alternatively, a asynchronous handshake can be adopted by using the bus strobe and wait pins (BS* and WT*).

Using a DSP as a microcontroller is becoming another common trend. The processor has memory and peripherals which makes it look like a microcontroller — albeit one with a very fast processing capability and slightly different programming techniques. This, coupled with the increasing need for some form of DSP function such as filtering in many embedded systems, has meant that DSP controllers are a feasible choice for embedded designs.

Choosing a processor

So far in this chapter, the main processor types used in embedded systems along with various examples have been discussed. There are very many types available ranging in cost, processing power and levels of integration. The question then arises concerning how do you select a processor for an embedded system?

The two graphs show the major trends with processors. The first plots system cost against performance. It shows that for the highest performance discrete processors are needed and these have the highest system cost. For the lowest cost, microcontrollers are the best option but they do not offer the level of performance that integrated or discrete processors offer. Many use the 8 bit accumulator processor architecture which has now been around for over 20 years. In between the two are the integrated processors which offer medium performance with medium system cost.

The second graph shows the trend towards system integration against performance. Microcontrollers are the most integrated, but as stated previously, they do not offer the best performance. However, the ability to pack a whole system including memory, peripherals and processor into a single package is attractive, provided there is enough performance to perform the work required.

The problem comes with the overlap areas where it becomes hard to work out which way to move. This is where other factors come into play:

Does it have enough performance?

A simple question to pose but a difficult one to answer. The problem is in defining the type of performance that is needed. It may be the ability to perform integer or floating point arithmetic operations or the ability to move data from one location to another. Another option may be the interrupt response time to allow data to be collected or processed.

The problem is that unless the end system is understood it is difficult to know exactly how much performance is needed. Add to this the uncertainty in the software programming overhead, i.e. the performance loss in using a high level language compared to low level assembler, and it is easy to see why the answer is not straightforward.

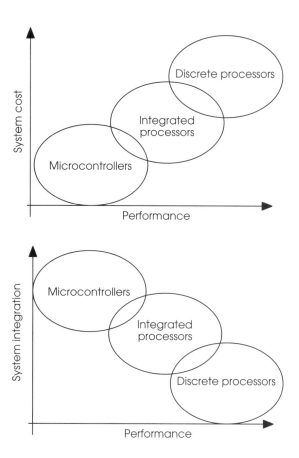

Processor selection graphs

In practice, most paper designs assume that about 20–40% of the processor performance will be lost to overheads in terms of MIPs and processing. Interrupt latencies can be calculated to give more accurate figures but as will be explained in Chapter 7, this has its own set of problems and issues to consider.

This topic of selecting and configuring a processor is discussed in many of the design notes and tutorials at the end of this book.

3   Memory systems

Within any embedded system, memory is an important part of the design, and faced with the vast variety of memory that is available today, choosing and selecting the right type for the right application is of paramount importance. This chapter goes through the different types that are available and discusses the issues associated with them that influence the design.

Memory technologies

Within any embedded system design that uses external memory, it is almost a sure bet that the system will contain a mixture of non-volatile memory such as EPROM (erasable programmable read only memory) to store the system software and DRAM (dynamic random access memory) for use as data and additional program storage. With very fast systems, SRAM (static random access memory) is often used as a replacement for DRAM because of its faster speed or within cache memory subsystems to help improve the system speed offered by DRAM.

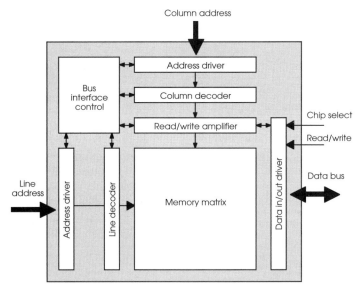

RAM block diagram

The main signals used with memory chips fall into several groups:

- Address bus
  The address bus is used to select the particular location within the memory chip. The signals may be multiplexed as in the case with DRAM or non-multiplexed as with SRAM.

- Data bus

  This bus provides the data to and from the chip. In some cases, the memory chip will use separate pins for incoming and outgoing data, but in others a single set of pins is used with the data direction controlled by the status of chip select signals, the read/write pin and output enable pins.

- Chip selects

  These can be considered as additional address pins that are used to select a specific chip within an array of memory devices. The address signals that are used for the chip selects are normally the higher order pins. In the example shown, the address decode logic has enabled the chip select for the second RAM chip — as shown by the black arrow — and it is therefore the only chip driving the data bus and supplying the data. As a result, each RAM chip is located in its own space within the memory map although it shares the same address bus signals with all the other RAM chips in the array.

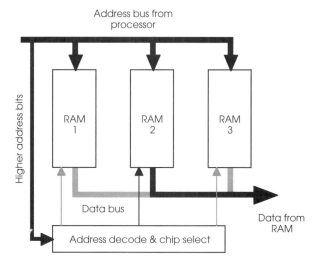

Address decode and chip select generation

- Control signals including read/write signals

  Depending on the functionality provided by the memory device, there are often additional control signals. Random access memory will have a read/write signal to indicate to type of access. This is missing from read only devices such as EPROM. For devices that have multiplexed address buses, as with the case with DRAM, there are control signals associated with this type of operation.

  There are now several different types of semiconductor memory available which use different storage methods and have different interfaces.

DRAM technology

DRAM is the predominantly used memory technology for PCs and embedded systems where large amounts of low cost memory are needed. With most memory technologies, the cost per bit is dependent on two factors: the number of transistors that are used to store each bit of data and the type of package that is used. DRAM achieves its higher density and lower cost because it only uses a single transistor cell to store each bit of data. The data storage element is actually a small capacitor whose voltage represents a binary zero or one which is buffered by the transistor. In comparison, a SRAM cell contains at least four or five transistors to store a single bit of data and does not use a capacitor as the active storage element. Instead, the transistors are arranged to form a flip-flop logic gate which can be flipped from one binary state to the other to store a binary bit.

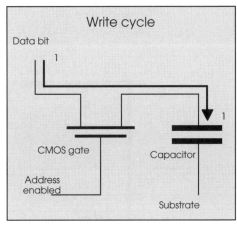

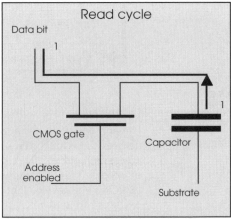

DRAM cell read and write
cycles

DRAM technology does have its drawbacks with the major one being its need to be refreshed on a regular basis. The term 'dynamic' refers to the memory's constant need for its data to be refreshed. The reason for this is that each bit of data is stored using a capacitor, which gradually loses its charge. Unless it is frequently topped up (or refreshed), the data disappears.

This may appear to be a stupid type of memory — but the advantage it offers is simple — it takes only one transistor to store a bit of data whereas static memory takes four or five. The memory chip's capacity is dependent on the number of transistors that can be fabricated on the silicon and so DRAM offers about four times the storage capacity of SRAM (static RAM). The refresh overhead takes about 3–4% of the theoretical maximum processing available and is a small price to pay for the larger storage capacity. The refresh is performed automatically either by a hardware controller or through the use of software. These techniques will be described in more detail later on in this chapter.

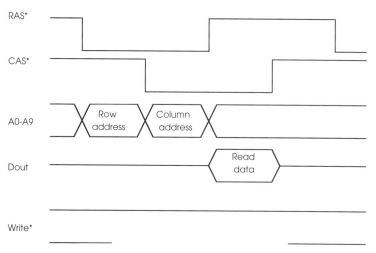

Basic DRAM interface

The basic DRAM interface takes the processor generated address, places half of the address (the high order bits) onto the memory address bus to form the row address and asserts the RAS* signal. This partial address is latched internally by the DRAM. The remaining half (the low order bits), forming the column address, are then driven onto the bus and the CAS* signal asserted. After the access time has expired, the data appears on the Dout pin and is latched by the processor. The RAS* and CAS* signals are then negated. This cycle is repeated for every access. The majority of DRAM specifications define minimum pulse widths for the RAS* and CAS* and these often form the major part in defining the memory access time. When access times are quoted, they usually refer to the time from the assertion of the RAS* signal to the appearance of the data. There are several variations on this type of interface, such as page mode and EDO. These will be explained later on in this chapter

Video RAM

A derivative of DRAM is the VRAM (Video RAM), which is essentially a DRAM with the ability for the processor to update its contents at the same time as the video hardware uses the data to create the display. This is typically done by adding a large shift register to a normal DRAM. This register can be loaded with a row or larger amounts of data which can then be serially clocked out to the video display. This operation is in parallel with normal read/write operations using a standard address/data interface.

SRAM

SRAM does not need to be refreshed and will retain data indefinitely — as long as it is powered up. In addition it can be designed to support low power operation and is often used in preference to DRAM for this reason. Although the SRAM cell contains more transistors, the cell only uses power when it is being switched. If the cell is not accessed then the quiescent current is extremely low. DRAM on the other hand has to be refreshed by external bus accesses and these consume a lot of power. As a result, the DRAM memory will have a far higher quiescent current than that of SRAM.

The SRAM memory interface is far simpler than that of DRAM and consists of a non-multiplexed address bus and data bus. There is normally a chip select pin which is driven from other address pins to select a particular SRAM when they are used in banks to provide a larger amount of storage.

Typical uses for SRAM include building cache memories for very fast processors, using as main memory in portable equipment where its lower power consumption is important and as expansion memory for microcontrollers.

Pseudo-static RAM

Pseudo-static RAM is a memory chip that uses DRAM cells to provide a higher memory density but has the refresh control built into the chip and therefore acts like a static RAM. It has been used in portable PCs as an alternative to SRAM because of its low cost.

Battery backed-up SRAM

The low power consumption of SRAM makes it suitable for conversion into non-volatile memories, i.e. memory that does not lose its data when the main power is removed by adding a small battery to provide power at all times. With the low quiescent current often being less than the battery's own leakage current, the SRAM can be treated as a non-volatile RAM for the duration of the battery's life. The CMOS (complementary metal oxide semiconductor) memory used by the MAC and IBM PC, which contains the configuration data, is SRAM. It is battery backed up to ensure it is powered up while the computer is switched off.

Some microcontrollers with on-chip SRAM support the connection of an external battery to backup the SRAM contents when the main power is removed.

EPROM and OTP

EPROM is used to store information such as programs and data that must be retained when the system is switched off. It is used within PCs to store the Toolbox and BIOS routines and power on software in the MAC and IBM PC respectively that is executed when the computer is switched on. These devices are read only and cannot be written to, although they can be erased by ultraviolet (UV) light and have a transparent window in their case for this purpose. This window is usually covered with a label to prevent accidental erasure, although it takes 15–30 minutes of intense exposure to do so.

There is a different packaged version of EPROM called OTP (one time programmable) which is an EPROM device packaged in a low cost plastic package. It can be programmed once only because there is no window to allow UV light to erase the EPROM inside. These are becoming very popular for small production runs, for example.

Flash

Flash memory is a non-volatile memory which is electrically erasable and offers access times and densities similar to that of DRAM. It uses a single transistor as a storage cell and by placing a higher enough charge to punch through an oxide layer, the transistor cell can be programmed. This type of write operation can take several milliseconds compared to sub 100 ns for DRAM or faster for SRAM. Reads are of the order of 70–100 ns.

FLASH has been positioned and is gaining ground as a replacement for EPROMs but it has not succeeded in replacing hard disk drives as a general purpose form of mass storage. This is due to the strides in disk drive technology and the relatively slow write access time and the wearout mechanism which limits the number of writes that can be performed. Having said this, it is frequently used in embedded systems that may need remote software updating. A good example of this is with modems where the embedded software is stored in FLASH and can be upgraded by downloading the new version via the Internet or bulletin board using the modem itself. Once downloaded, the new version can be transferred from the PC to the modem via the serial link.

EEPROM

Electrically erasable programmable read only memory is another non-volatile memory technology that is erased by applying a suitable electrical voltage to the device. These types of memory do not have a window to allow UV light in to erase them and thus offer the benefits of plastic packaging, i.e. low cost with the ability to erase and re-program many times.

The erase/write cycle is slow and can typically only be performed on large blocks of memory instead of at the bit or byte level. The erase voltage is often generated internally by a charge pump but can be supplied externally. The write cycles do have a wearout mechanism and therefore the memory may only be guaranteed for a few hundred thousand erase/write cycles and this, coupled with the slow access time, means that they are not a direct replacement for DRAM.

Memory organisation

A memory's organisation refers to how the data is arranged within the memory chips and within the array of chips that is used to form the system memory. An individual memory's storage is measured in bits but can be organised in several different ways. A 1 Mbit memory can be available as a 1 Mbit × 1 device, where there is only a single data line and eight are needed in parallel to store one byte of data. Alternatives are the 256 kbits × 4, where there are four data lines and only two are needed to store a byte, and 128 kbit × 8, which has 8 data lines. The importance of these different organisations becomes apparent when upgrading memory and determining how many chips are needed.

The minimum number of chips that can be used is determined by the width of the data path from the processor and the number of data lines the memory chip has. For an MC68000 processor with a 16 bit wide data path, 16 × 1 devices, four × 4 or two × 8 devices would be needed. For a 32 bit processor, like the MC68020, MC68030, MC68040, 80386DX or 80486, this figure doubles. What is interesting is that the wider the individual memory chip's data storage, the smaller the number of chips that is required to upgrade. This does not mean that, for a given amount of memory, less × 4 and × 8 chips are needed when compared with × 1 devices, but that each minimum upgrade can be smaller, use fewer chips and be less expensive. With a 32 bit processor and using 1 Mbit × 1 devices, the minimum upgrade would need 32 chips and add 32 Mbytes. With a × 4 device, the minimum upgrade would only need 8 chips and add 8 Mbytes.

This is becoming a major problem as memories become denser and the smaller size chips are discontinued. This poses problems to designers that need to design some level of upgrade capability to cater for the possible — some would say inevitable — need for more memory to store the software. With the *smallest* DRAM chip that is still in production being a 1 Mbit device and the likelihood that this will be replaced by 4 and 16 Mbit devices in the not so distant future, the need for one additional byte could result in the addition of 1 or 4 Mbytes or memory. More importantly, if a × 1 organisation is used, then this means that an additional 8 chips are needed. By using a wider organisation, the number of chips is reduced.

By 1 organisation

Today, single-bit memories are not as useful as they used to be and their use is in decline compared to wider data path devices. Their use is restricted to applications that need non-standard width memory arrays that these type of machines use, e.g. 12 bit, 17 bit etc. They are still used to provide a parity bit and can be found on SIMM memory modules but as systems move away from implementing parity memory — many PC motherboards no longer do so — the need for such devices will decline.

By 4 organisation

This configuration has effectively replaced the × 1 memory in microprocessor applications because of its reduced address bus loading and complexity — only 8 chips are needed to build a 32 bit wide data path instead of 32 and only two are needed for an 8 bit wide bus.

By 8 and by 9 organisations

Wider memories such as the × 8 and × 9 are beginning to replace the × 4 parts in many applications. Apart from higher integration, there are further reductions in address bus capacitance to build a 32 or 64 bit wide memory array. The reduction in bus loading can improve the overall access time by greatly reducing the address set-up and stabilisation time, thus allowing more time within the memory cycle to access the data from the memories. This improvement can either reduce costs by using slower and cheaper memory, or allow a system to run faster given a specific memory part. The × 9 variant provides a ninth bit for parity protection. For microcontrollers, these parts allow memory to be increased in smaller increments.

By 16 and greater organisations

Wider memories with support for 16 bits or wider memory are already appearing but it is likely that they will integrate more of the interface logic so that the time consumed by latches and buffers during the memory access will be removed, thus allowing slower parts to be used in wait state-free designs.

Parity

The term parity has been mentioned in the previous paragraphs along with statements that certainly within the PC industry it is no longer mandatory and the trend is moving away from its implementations. Parity protection is an additional bit of memory which is used to detect single-bit errors with a block of memory. Typically, one parity bit is used per byte of data. The bit is set to a one or a zero depending on the number of bits that are set to one within the data byte. If this number is odd, the parity bit is set to a one and if the number is even, it is set to zero. This can be reversed to provide two parity schemes known as odd and even parity.

If a bit is changed within the word through an error or fault, then the parity bit will no longer be correct and a comparison of the parity bit and the calculated parity value from the newly read data will disagree. This can then be used to flag a signal back to the processor, such as an error. Note that parity does not allow the error to be corrected nor does it protect from all multiple bit failures such as two set or cleared bits failing together. In addition it requires a parity controller to calculate the value of the parity bit on write cycles and calculate and compare on read cycles. This additional work can slow down memory access and thus the processor performance.

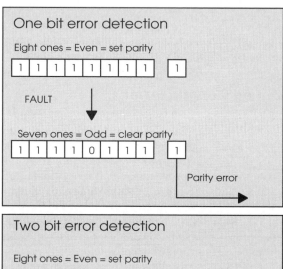

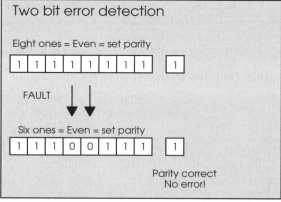

Parity detection for one and two bit errors

However, for critical embedded systems it is important to know if there has been a memory fault and parity protection may be a requirement.

Parity initialisation

If parity is used, then it may be necessary for software routines to write to each memory location to clear and/or set up the parity hardware. If this is not done, then it is possible to generate false parity errors.

Error detecting and correcting memory

With systems that need very high reliability, it is possible through increasing the number of additional bits per byte and by using special coding techniques to increase the protection offered by parity protection. There are two types of memory design that do this:

- Error detecting memory

    With this type of memory, additional bits are added to the data word to provide protection from multiple bit failures. Depending on the number of bits that are used and the coding techniques that are use the additional bits, protection can provided for a larger number of error conditions. The disadvantage is of the additional memory bits that are needed along with the complex controllers needed to create and compare the codes.

- Error detecting and correction

    This takes the previous protection one step further and uses the codes not only to detect the error but correct it as well. This means that the system will carry on despite the error whereas the previous scheme would require the system to be shut down as it could not rely on the data. EDC systems, as they are known, are expensive but offer the best protection against memory errors.

Access times

As well as different sizes and organisations, memory chips have different access times. The access time is the maximum time taken by the chip to read or write data and it is important to match the access time to the design. (It usually forms part of the part number: MCM51000AP10 would be a 100 ns access time memory and MCM51000AP80 would be an 80 ns version.) If the chip is too slow, the data that the processor sees will be invalid and corrupt, resulting in software problems and crashes. Some designs allow memories of different speed to be used by inserting wait states between the processor and memory so that sufficient time is given to allow the correct data to be obtained. These often require jumper settings to be changed or special set-up software to be run, and depend on the manufacture and design of the board.

If a processor clock speed is increased, the maximum memory access time must be reduced — so changing to a faster processor may require these settings to be modified.

Packages

The major option with memories is packaging. Some come encapsulated in plastic, some in a ceramic shell and so on. There are many different types of package options available and, obviously, the package must match the sockets on the board. Of the many different types, four are commonly used: the dual in line package, zig–zag package, SIMM and SIP. The most common package encountered with the MAC is the SIMM, although all the others are used, especially with third party products.

Dual in line package

This package style, as its name implies, consists of two lines of legs either side of a plastic or ceramic body. It is the most commonly used package for the BIOS EPROMs, DRAM and SRAM. It is available in a variety of sizes with 24, 26 and 28 pin packages used for EPROMs and SRAMs and 18 and 20 pin packages for 1 Mbit × 1 and 256 kbit × 4 DRAMs. However, it has virtually been replaced by the use of SIMM modules and is only used for DRAM on the original MAC 128K and 512K models for DRAM and for EPROM on models up to the MAC IIx.

Zig–zag package

This is a plastic package used primarily for DRAM. Instead of coming out of the sides of the package, the leads protrude from the pattern and are arranged in a zig-zag — hence the name. This type of package can be more difficult to obtain, compared with the dual in line devices, and can therefore be a little more expensive. This format is often used on third party boards.

SIMM

SIMM is not strictly a package but a subassembly. It is a small board with finger connection on the bottom and sufficient memory chips on board to make up the required configuration, such as 256 Kbit × 8 or × 9, 1 Mbit × 8 or × 9, 4 Mbit, and so on. SIMMs have rapidly gained favour and many new designs use these boards instead of individual memory chips. They require special sockets, which can be a little fragile and need to be handled correctly. There are currently two types used for PCs: the older 30 pin SIMM which uses an 8 or 9 bit (8 bits plus a parity bit) data bus and a more recent 72 pin SIMM which has a 32 or 36 bit wide data bus. The 36 bit version is 32 bits data plus four parity bits. Apple has used both types and a third which has 64 pins but like the IBM PC world standardised on the 72 pin variety.

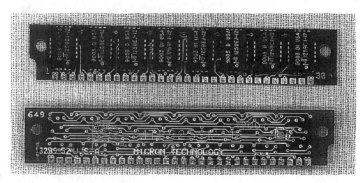

256 kbyte 30 pin SIMM

The older 30 pin SIMMs are normally used in pairs for a 16 bit processor bus (80386SX, MC68000) and in fours for 32 bit buses (80386DX, 80486, MC68030, MC68040) while the 72 pin SIMMS are normally added singly although some higher performance boards need a pair of 72 pin SIMMs to support bank switching.

As a result of this increased bus bandwidth, 168 pin SIMMs known as DIMMs are increasingly being used with high performance systems.

While the common use is for DRAM, Apple also use SIMM modules for video RAM and the Toolbox ROMs.

SIPP

This is the same idea as SIMM, except that the finger connections are replaced by a single row of pins. SIPP has been overtaken by SIMM in terms of popularity and is now rarely seen.

DRAM interfaces

The basic DRAM interface

The basic DRAM interface takes the processor generated address, places the high order bits onto the memory address bus to form the row address and asserts the RAS* signal. This partial address is latched internally by the DRAM. The remaining low order bits, forming the column address, are then driven onto the bus and the CAS* signal asserted. After the access time has expired, the data appears on the Dout pin and is latched by the processor. The RAS* and CAS* signals are then negated. This cycle is repeated for every access. The majority of DRAM specifications define minimum pulse widths for the RAS* and CAS* and these often form the major part in defining the memory access time. To remain compatible with the PC–AT standard, memory refresh is performed every 15 microseconds.

This direct access method limits wait state-free operation to the lower processor speeds. DRAM with 100 ns access time would only allow a 12.5 MHz processor to run with zero wait states. To achieve 20 MHz operation needs 40 ns DRAM, which is unavailable today, or fast static RAM which is at a price. Fortunately, the embedded system designer has more tricks up his sleeve to improve DRAM performance for systems, with or without cache.

Page mode operation

One way of reducing the effective access time is to remove the RAS* pulse width every time the DRAM was accessed. It needs to be pulsed on the first access, but subsequent accesses to the same page (i.e. with the same row address) would not require it and so are accessed faster. This is how the 'page mode' versions of most 256 kb, 1 Mb and 4 Mb memory work. In page mode, the row address is supplied as normal but the RAS* signal is left asserted. This selects an internal page of memory within the DRAM where any bit of data can be accessed by placing the column address and asserting CAS*. With 256 kb size memory, this gives a page of 1 kbyte (512 column bits per DRAM row × 16 DRAMs in the array). A 2 kbyte page is available from 1 Mb DRAM and a 4 kbyte page with 4 Mb DRAM.

This allows fast processors to work with slower memory and yet achieve almost wait state-free operation. The first access is slower and causes wait states but subsequent accesses within the selected page are quicker with no wait states.

However, there is one restriction. The maximum time that the RAS* signal can be asserted during page mode operation is often specified at about 10 microseconds. In non-PC designs, the refresh interval is frequently adjusted to match this time, so a refresh cycle will always occur and prevents a specification violation. With the PC standard of 15 microseconds, this is not possible. Many chip sets neatly resolve the situation by using an internal counter which times out page mode access after 10 microseconds.

Page interleaving

Using a page mode design only provides greater performance when the memory cycles exhibit some form of locality, i.e. stay within the page boundary. Every access outside the boundary causes a page miss and two or three wait states. The secret, as with caches, is to increase the hits and reduce the misses. Fortunately, most accesses are sequential or localised, as in program subroutines and some data structures. However, if a program is frequently accessing data, the memory activity often follows a code–data–code–data access pattern. If the code areas and data areas are in different pages, any benefit that page mode could offer is lost. Each access changes the page selection, incurring wait states. The solution is to increase the number of pages available. If the memory is divided into several banks, each bank can offer a selected page, increasing the number of pages and, ultimately, the number of hits and performance. Again, extensive hardware support is needed and is frequently provided by the PC chip set.

Page interleaving is usually implemented as a one, two or four way system, depending on how much memory is installed. With a four way system, there are four memory banks, each with their own RAS* and CAS* lines. With 4 Mbyte DRAM, this would offer 16 Mbytes of system RAM. The four way system allows four pages to be selected within page mode at any one time. Page 0 is in bank 1, page 1 in bank 2, and so on, with the sequence restarting after four banks.

With interleaving and Fast Page Mode devices, inexpensive 85 ns DRAM can be used with a 16 MHz processor to achieve a 0.4 wait state system today. With no page mode interleaving, this system would insert two wait states on every access. With the promise of faster DRAM, future systems will be able to offer 33–50 MHz with very good performance — without the need for cache memory and its associated costs and complexity.

Burst mode operation

Some versions of the DRAM chip, such as page mode, static column or nibble mode devices, do not need to have the RAS/CAS cycle repeated and can provide data much faster if only the new column address is given. This has allowed the use of a burst fill memory interface, where the processor fetches more data than it needs and keeps the extra data in an internal cache ready for future use. The main advantage of this system is in reducing the need for fast static RAMs to realise the processor's performance. With 60 ns page mode DRAM, a 4-1-1-1 (four clocks for the first access, single cycle for the remaining burst) memory system can easily be built. Each 128 bits of data fetched in such a way takes only 7 clock cycles, compared with 5 in the fastest possible system. If bursting was not supported, the same access would take 16 clocks. This translates to a very effective price performance — a 4-1-1-1 DRAM system gives about 90% of the performance of a more expensive 2-1-1-1 static RAM design. This interface is used on the higher performance processors where it is used in conjunction with on-chip caches. The burst fill is used to load a complete line of data within the cache.

This allows fast processors to work with slower memory and yet achieve almost wait state-free operation. The first access is slower and causes wait states but subsequent accesses within the selected page are quicker with no wait states.

EDO memory

EDO stands for extended data out memory and is a form of fast page mode RAM that has a quicker cycling process and thus faster page mode access. This removes wait states and thus improves the overall performance of the system. The improvement is achieved by fine tuning the CAS* operation.

With fast page mode when the RAS* signal is still asserted, each time the CAS* signal goes high the data outputs stop asserting the data bus and go into a high impedance mode. This is used to simplify the design by using this transistion to meet the timing requirements. It is common with this type of design to permanently ground the output enable pin, for example. The problem is that this requires the CAS* signal to be asserted until the data from the DRAM is latched by the processor or bus master. These means that the next access cannot be started until this has been completed, causing delays.

EDO memory do not cause the outputs to go high impedance and it will continue to drive data even if the CAS* signal is removed. By doing this, the CAS* precharge can be started for the next access while the data from the previous access is being latched. This saves valuable nanoseconds and can mean the removal of a wait state. With very high performance processors, this is a big advantage and EDO type DRAM is becoming the *de facto* standard for PCs and workstations or any other application that needs high performance memory.

DRAM refresh techniques

DRAM needs to be periodically refreshed and to do this there are several methods that can be used. The basic technique involves accessing the DRAM using a special refresh cycle. During these refresh cycles, no other access is permitted. The whole chip must be refreshed within a certain time period or its data will be lost. This time period is known as the refresh time. The number of accesses needed to complete the refresh is known as the number of cycles and this number divided into the refresh time gives the refresh rate. There are two refresh rates in common use: standard which is 15.6 µs and extended which is 125 µs. Each refresh cycle is approximately twice the length of a normal access — a 70 ns DRAM typically as a refresh cycle time of 130 ns — and this times the number of cycles gives the total amount of time lost in the refresh time to refresh. This figure is typically 3–4% of the refresh time. During this period, the memory is not accessible and thus any processor will have to wait for its data. This raises some interesting potential timing problems.

Distributed vs. burst refresh

With a real–time embedded system, the time lost to refresh must be accounted for. However, its effect is dependent on the method chosen to perform all the refresh cycles within the refresh time. A 4 M by 1 DRAM requires 1024 refresh cycles. Are these cycles executed in a burst all at once or should they be distributed across the whole time. Bursting means that the worst case delay is 1024 times larger than that of a single refresh cycle that would be encountered in a distributed system. This delay is of the order of 0.2 ms, a not inconsiderable time for many embedded systems! The distributed worst case delay due to refresh is about 170 ns.

Most systems use the distributed method and depending on the size of time critical code, calculate the number of refresh cycles that are likely to be encountered and use that to estimate the delay caused by refresh cycles. It should be remembered that in both cases, the time and access overhead for refresh is the same.

Software refresh

It is possible to use software to perform the refresh by using a special routine to periodically circle through the memory and thus cause its refresh. Typically a timer is programmed to generate an interrupt. The interrupt handler would then perform the refresh. The problem with this arrangement is that any delay in performing the refresh potentially places the whole memory and its contents at risk. If the processor is stopped or single stepped, its interrupts disabled or similar, the refresh is halted and the memory contents lost. The disadvantage in this is that it makes debugging such a system extremely difficult. Many of the debugging techniques cannot be used because they stop the refresh. If the processor crashes, the refresh is stopped and the contents are lost.

There have been some neat applications where software refresh is used. The Apple II personal computer used a special memory configuration so that every time the DRAM blocks that were used for video memory were accessed to update the screen, they effectively refreshed the DRAM.

RAS only refresh

With this method, the row address is placed on the address bus, RAS* is asserted but CAS* is held off. This generates the recycle address. The address generation is normally done by an external hardware controller, although many early controllers required some software assistance. The addressing order is not important but what is essential is that all the rows are refreshed within the refresh time.

CAS before RAS (CBR) refresh

This is a later refresh technique that is now commonly used. It has lower power consumption because it does not use the address bus and the buffers can be switched off. It works by using an internal address counter stored on the memory chip itself which is periodically incremented. Each incrementation starts a refresh cycle internally. The mechanism works as its name suggests by asserting CAS* before RAS*. Each time that RAS* is asserted, a refresh cycle is performed and the internal counter incremented.

Hidden refresh

This is a technique where a refresh cycle is added to the end of a normal read cycle. The term hidden refers to the fact that the refresh cycle is hidden in a normal read and not to any hiding of the refresh timing. It does not matter which technique you use, refresh will still cost time and performance! What happens is that the RAS* signal goes high and then is asserted low. This happens at the end of the read cycle when the CAS* signal is still asserted. This is a similar situation to the CBR method. Like it, this toggling of the RAS* signal at the end of the read cycle starts a CBR refresh cycle internally.

Cache memory

With faster (>25 MHz) processors, the wait states incurred in DRAM accesses start to dramatically reduce performance. To recover this, many designs implement a cache memory to buffer the processor from such delays.

Cache memory systems work because of the cyclical structures within software. Most software structures are loops where pieces of code are repeatedly executed, albeit with different data. Cache memory systems store these loops so that after the loop has been fetched from main memory, it can be obtained from the cache for subsequent executions. The accesses from cache are faster than from main memory and thus increase the system's throughput.

There are several criteria associated with cache design which affect its performance. The most obvious is cache size — the larger the cache, the more entries are stored and the higher the hit rate. For the 80x86 processor architecture, the best price performance is obtained with a 64 kbyte cache. Beyond this size, the cost of getting extra performance is extremely high.

The set associativity is another criterion. It describes the number of cache entries that could possibly contain the required data. With a direct map cache, there is only a single possibility, with a two way system, there are two possibilities, and so on. Direct mapped caches can get involved in a bus thrashing situation, where two memory locations are separated by a multiple of the cache size. Here, every time word A is accessed, word B is discarded from the cache. Every time word B is accessed, word A is lost, and so on. The cache starts thrashing and overall performance is degraded. With a two way design, there are two possibilities and this prevents bus thrashing. The cache line refers to the number of consecutive bytes that are associated with each cache entry. Due to the sequential nature of instruction flow, if a cache hit occurs at the beginning of the line, it is highly probable that the rest of the line will be accessed as well. It is therefore prudent to burst fill cache lines whenever a miss forces a main memory access. The differences between set associativity and line length are not as clear as cache size. It is difficult to say what the best values are for a particular system. Cache performances are extremely system and software dependent and, in practice, system performance increases of 20–30% are typical.

Cache coherency

The biggest challenge with cache design is how to solve the problem of data coherency, while remaining hardware and software compatible. The issue arises when data is cached which can then be modified by more than one source. An everyday analogy is that of a businessman with two diaries — one kept by his secretary in the office and the other kept by him. If he is out of the office and makes an appointment, the diary in the office is no longer valid and his secretary can double book him assuming, incorrectly, that the office diary is correct.

This problem is normally only associated with data but can occur with instructions within an embedded application. The stale data arises when a copy is held both in cache and in main memory. If either copy is modified, the other becomes stale and system coherency is destroyed. Any changes made by the processor can be forced through to the main memory by a 'write through' policy, where all writes automatically update cache and main memory. This is simple to implement but does couple the processor unnecessarily to the slow memory. More sophisticated techniques, like 'copy back' and 'modified write back' can give more performance (typically 15%, although this is system and software dependent) but require bus snooping support to detect accesses to the main memory when the valid data is in the cache.

The 'write through' mechanism solves the problem from the processor perspective but does not solve it from the other direction. DMA (Direct Memory Access) can modify memory directly without any processor intervention. Consider a task swapping system. Task A is in physical memory and is cached. A swap occurs and task A is swapped out to disk and replaced by task B at the same location. The cached data is now stale. A software solution to this involves flushing the cache when the page fault happens so the previous contents are removed. This can destroy useful cached data and needs operating system support, which can make it non–compatible. The only hardware solution is to force any access to the main memory via the cache, so that the cache can update any modifications.

This provides a transparent solution — but it does force the processor to compete with the DMA channels and restricts caching to main memory only, with a resultant impact on performance.

While many system designs use cache memory to buffer the fast processor from the slower system memory, it should be remembered that access to system memory is needed on the first execution of an instruction or software loop and whenever a cache miss occurs. If this access is too slow, these overheads greatly diminish the efficiency of the cache and, ultimately, the processor's performance. In addition, switching on caches can cause software that works perfectly to crash and, in many configurations, the caches remain switched off to allow older software to execute correctly.

A lot is made of cache implementations — but unless the main system memory is fast and software reliable, system and software performance will degrade. Caches help to regain performance lost through system memory wait states but they are never 100% efficient. A system with no wait states always provides the best performance.

Burst interfaces

The adoption of burst interfaces by virtually all of today's high performance processors has led to the development of special memory interfaces which include special address generation and data latches to help the designer. Burst interfaces take advantage of page and nibble mode memories which supply data on the first access in the normal time, but can supply subsequent data far quicker.

The burst interface, which is used on processors from Motorola, Intel, AMD, MIPs and many other manufacturers gains its performance by fetching data from memory in bursts from a line of sequential locations. It makes use of a burst fill technique where the processor will access typically four words in succession, enabling a complete cache line to be fetched or written out to memory. The improved speed is obtained by taking advantage of page mode or static column memory. These type of memories offer faster access times — single cycle in many cases — after the initial access is made. The advantage is a reduction in clock cycles needed to fetch the same amount of data. To fetch four words with a three clock memory cycle takes 12 clocks. Fetching the same amount of data using a 2-1-1-1 burst (two clocks for the first access, single cycle for the remainder) takes only five clocks. This type of interface gives a good fit with the page mode DRAM where the first access is used to set up the page access and the remainder of the burst accesses addresses within the page, thus taking advantage of the faster access.

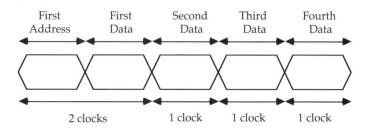

| First Address | First Data | Second Data | Third Data | Fourth Data |

2 clocks   1 clock   1 clock   1 clock

A burst fill interface

Burst fill offers advantages of faster and more efficient memory accesses, but there are some fundamental changes in its operation when compared with single access buses. This is particularly so with SRAM when it is used as part of a cache:

- The address is only supplied for the first access in a burst and not for the remaining accesses. External logic is required to generate the additional addresses for the memory interface.

- The timing for each data access in the burst sequence is unequal: typical clock timings are 2-1-1-1 where two clocks are taken for the first access, but subsequent accesses are single cycle.

- The subsequent single cycle accesses compress address generation, set up and hold and data access into a single cycle, which can cause complications generating write pulses to write data into the SRAM, for example.

These characteristics lead to conflicting criteria within the interface: during a read cycle, the address generation logic needs to change the address to meet set-up and hold times for the next access, while the current cycle requires the address to remain constant during its read access. With a write cycle, the need to change the address for the next cycle conflicts with the write pulse and constant address required for a successful write.

Meeting the interface needs

For a designer implementing such a system there are four methods of improving the SRAM interface and specification to meet the timing criteria:

- Use faster memory.
- Use synchronous memory with onchip latches to reduce gate delays.
- Choose parts with short write pulse requirements and data set-up times.
- Integrate address logic on chip to remove the delays and give more time.

While faster and faster memories are becoming available, they are more expensive, and memory speeds are now becoming limited by on- and off-chip buffer delays rather than the cell access times. The latter three methods depend on semiconductor manufacturers recognising the designer's difficulties and providing static RAMs which interface better with today's high performance processors.

This approach is beneficially for many high speed processors, but it is not a complete solution for the burst interfaces. They still need external logic to generate the cyclical addresses from the presented address at the beginning of the burst memory access. This increases the design complexity and forces the use of faster memories than normally necessary simply to cope with the propagation delays. The obvious step is to add this logic to the latches and registers of a synchronous memory to create a protocol specific memory that supports certain bus protocols. The first two members of Motorola's Protocol specific products are the MCM62940 and MCM62486 32K × 9 fast static RAMs. They are, as their part numbering suggests, designed to support the MC68040 and the Intel 80486 bus burst protocols. These parts offer access times of 15 and 20 ns.

The first access may take two processor clocks but remaining accesses can be made in a single cycle. There are some restrictions to this: the subsequent accesses must be in the same memory page and the processor must have somewhere to store the extra data that can be collected. The obvious solution is to use this burst interface to fill a cache line. The addresses will be in the same page and by storing the extra data in a cache allows a processor to use it at a later date without consuming additional bus bandwidth. The main problem faced by designers with these interfaces is the generation of the new addresses. In most designs the processor will only issue the first address and will hold this constant during the extra accesses. It is up to the interface logic to take this address and increment it with every access. With processors like the MC68030, this function is a straight incremental count. With the MC68040, a wrap-around burst is used where the required data is fetched first and the rest of the line fetched, wrapping around to the line beginning if necessary. Although more efficient for the processor, the wrap-around logic is more complicated.

The solution is to add this logic along with latches and registers to a memory to create a specific part that supports certain bus protocols. The first two members of Motorola's Protocol specific products are the MCM62940 and MCM62486 32K × 9 fast static RAMs.

They are, as their part numbering suggests, designed to support the MC68040 and the Intel 80486 bus burst protocols. These parts offer access times of 15 and 20 ns.

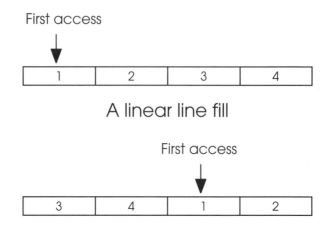

First access

| 1 | 2 | 3 | 4 |

A linear line fill

First access

| 3 | 4 | 1 | 2 |

A wrap-around line fill

Linear and wrap-around line fills

The MCM62940 has an on-chip burst counter that exactly matches the MC68040 wrap-around burst sequence. The address and other control data can be stored either by using the asynchronous or synchronous signals from the MC68040 depending on the design and its needs. A late write abort is supported which is useful in cache designs where cache writes can be aborted later in the cycle than normally expected, thus giving more time to decide whether the access should result in a cache hit or be delayed while stale data is copied back to the main system memory.

The MCM62486 has an on-chip burst counter that exactly matches the Intel 80486 burst sequence, again removing external logic and time delays and allowing the memory to respond to the processor without the need for the wait state normally inserted at the cycle start. In addition, it can switch from read to write mode while maintaining the address and count if a cache read miss occurs, allowing cache updating without restarting the whole cycle.

Big and little endian

There are two methods of organising data within memory depending on where the most significant bit is located. The Intel 80x86 and Motorola 680x0 and PowerPC processors use different organisations and this can cause problems.

The PowerPC architecture uses primarily a big endian byte order, i.e. an address points to the most significant byte of a value in memory. This can cause problems with other processors that use the alternative little endian organisation, where an address points to the least significant byte.

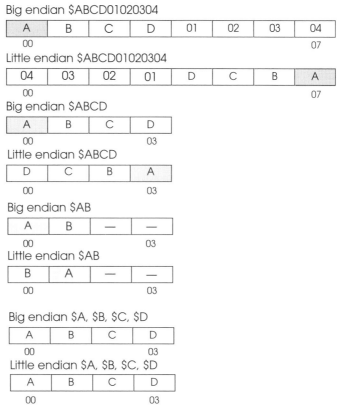

Big vs. little endian memory organisation

The PowerPC architecture solves this problem by providing a mode switch which causes all data memory references to be performed in little-endian fashion. This is done by swapping address bit lines instead of using data multiplexers. As a result, the byte swapping is not quite what may be expected and varies depending on the size of the data. It is important to remember that swapping the address bits only reorders the bytes and not the individual bits within the bytes. The bit order remains constant.

The diagram shows the different storage formats for big and little endian double words, words, half words and bytes. The most significant byte in each pair is shaded to highlight its position. Note that there is no difference when storing individual bytes.

An alternative solution for processors that do not implement the mode swapping is to use the load and store instructions that byte reverse the data as it moves from the processor to the memory and vice versa.

Dual port and shared memory

Dual port and shared memory are two types of memory that offer similar facilities, i.e. the ability of two processors to access the same memory and thus share data and/or programs. It is often used as a communication mechanism between processors. The difference between them concerns how they cope with two simultaneous accesses.

With dual port memory, such bus contention is resolved within the additional circuitry that is contained with the memory chip or interface circuitry. This usually consists of buffers that are used as temporary storage for one processor while the other accesses the memory. Both the memory accesses are completed as if there were only a single access.

The buffered information is transferred when the memory is available. If both accesses are writes to the same memory address, the first one to access the memory is normally given priority but this should not be assumed. Many systems consider this a programming error and use semaphores in conjunction with special test and set instructions to prevent this happening.

Shared memory resolves the bus contention by holding one of the processors off by inserting wait states into its memory access. This results in lost performance because the held off processor cannot do anything and has to wait for the other to complete. As a result, both processors lose performance because they are effectively sharing the same bus.

Shared memory is easier to design and is often used when large memory blocks are needed. Dual port memory is normally implemented with special hardware and is limited to relatively small memory blocks of a few kbytes.

Bank switching

Bank switching simply involves having several banks of memory with the same address locations. At any one time, only one bank of memory is enabled and accessible by the microprocessor. Bank selection is made by driving the required bank select line. These lines come from either an external parallel port or latch whose register(s) appear as a bank switching control register within the processors's normal address map.

In the example, the binary value 1000 has been loaded into the bank selection register. This asserts the select line for bank 'a' which is subsequently accessed by the processor. Special VLSI (very large scale integration) parts were even developed which provided large number of banks and additional control facilities: the Motorola MC6883 SAM is an example used in the Dragon MC6809-based home computer from the early 1980s.

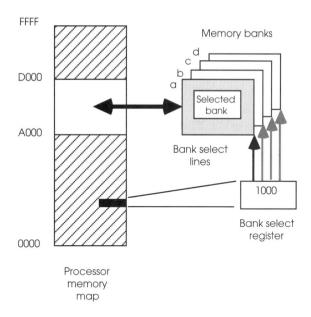

Bank switching

Memory overlays

Program overlays are similar to bank switching except that some form of mass storage is used to contain the different overlays. If a particular subroutine is not available, the software stores part of its memory as a file on disk and loads a new program section from disk into its place. Several hundred kilobytes of program, divided into smaller sections, can be made to overlay a single block of memory.

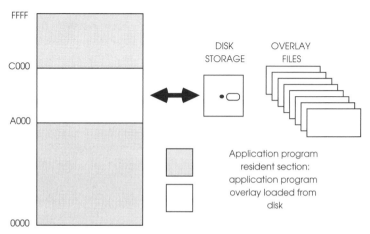

Program overlays, making a large program fit into the 64 kilobyte memory

This whole approach requires careful design so that the system integrity is ensured. Data passing between the overlays can be a particular problem which requires careful design. Typically, data is passed and stored either on the processor stack or in reserved memory which is locked in and does not play any part in the overlay process, i.e. it is resident all the time.

Example interfaces

MC68000 asynchronous bus

The MC68000 bus is fundamentally different to the buses used on the MC6800 and MC6809 processors. Their buses were synchronous in nature and assumed that both memory and peripherals could respond within a cycle of the bus. The biggest drawback with this arrangement concerned system upgrading and compatibility. If one component was uprated, the rest of the system needed uprating as well. It was for this reason that all M6800 parts had a system rating built into their part number. If a design specified an MC6809B, then it needed 2 MHz parts and subsequently, could not use an 'A' version which ran at 1 MHz. If a design based around the 1 MHz processor and peripherals was upgraded to 2 MHz, all the parts would need replacing. If a peripheral was not available at the higher speed, the system could not be upgraded. With the increasing processor and memory speeds, this restriction was unacceptable.

The MC68000 bus is truly asynchronous: it reads and writes data in response to inputs from memory or peripherals which may appear at any stage within the bus cycle. Provided certain signals meet certain set-up times and minimum pulse widths, the processor can talk to anything. As the bus is truly asynchronous it will wait indefinitely if no reply is received. This can cause similar symptoms to a hung processor; however, most system designs use a watchdog timer and the processor bus error signal to resolve this problem.

A typical bus cycle starts with the address, function codes and the read/write line appearing on the bus. Officially, this data is not valid until the address strobe signal AS* appears but many designs start decoding prior to its appearance and use the AS* to validate the output. The upper and lower data strobes, together with the address strobe signal (both shown as DS*), are asserted to indicate which bytes are being accessed on the bus. If the upper strobe is asserted, the upper byte is selected. If the lower strobe is asserted, the lower byte is chosen. If both are asserted together, a word is being accessed.

Once complete, the processor waits until a response appears from the memory or peripheral being accessed. If the rest of the system can respond without wait states (i.e. the decoding and access times will be ready on time) a DTACK* (Data Transfer ACKnowledge) signal is returned. This occurs slightly before clock edge S4. The data is driven onto the bus, latched and the address and data strobes removed to acknowledge the receipt of the DTACK* signal by the processor. The system responds by removing DTACK* and the cycle is complete. If the DTACK* signal is delayed for any reason, the processor will simply insert wait states into the cycle. This allows extra time for slow memory or peripherals to prepare data.

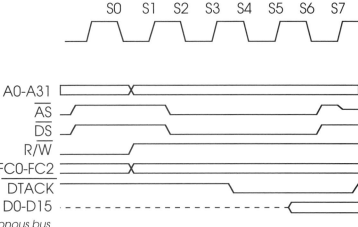

An MC68000 asynchronous bus cycle

The advantages that this offers are many fold. First, the processor can use a mixture of peripherals with different access speeds without any particular concern. A simple logic circuit can generate DTACK* signals with the appropriate delays as shown. If any part of the system is upgraded, it is a simple matter to adjust the DTACK* generation accordingly. Many M68000 boards provide jumper fields for this purpose and a single board and design can support processors running at 8, 10, 12 or 16 MHz. Secondly, this type of interface is very easy to interface to other buses and peripherals. Additional time can be provided to allow signal translation and conversion.

M6800 synchronous bus

Support for the M6800 synchronous bus initially offered early M68000 system designers access to the M6800 peripherals and allowed them to build designs as soon as the processor was available. With today's range of peripherals with specific M68000 interfaces, this interface is less used. However, the M6800 parts are now extremely inexpensive and are often used in cost-sensitive applications.

The additional signals involved are the E clock, valid memory address VMA* and valid peripheral address VPA*. The cycle starts in a similar way to the M68000 asynchronous interface except that DTACK* is not returned. The address decoding generates a peripheral chip select which asserts VPA*. This tells the M68000 that a synchronous cycle is being performed.

The address decoding monitors the E clock signal, which is derived from the main system clock, but is divided down by 10 with a 6:4 mark/space ratio. It is not referenced from any other signal and is free running. At the appropriate time (i.e. when E goes low) VMA* is asserted. The peripheral waits for E to go high and transfers the data. When E goes low, the processor negates VMA* and the address and data strobes to end the cycle.

For systems running at 10 MHz or lower, standard 1 MHz M6800 parts can be used. For higher speeds, 1.5 or 2 MHz versions must be employed. However, higher speed parts running at a lower clock frequency will not perform the peripheral functions at full performance.

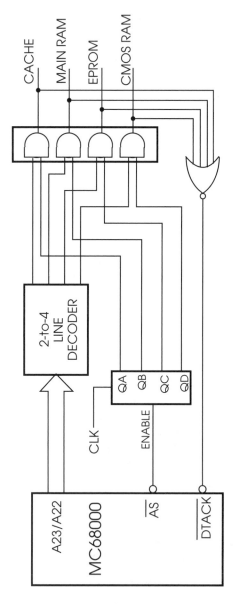

Example DTACK* generation

The MC68040 burst interface

Earlier in this chapter, some of the problems faced by a designer using SRAM with a burst interface were discussed. The MC68040 burst interface shows how these conflicts arise and their solution. It operates on a 2-1-1-1 basis, where two clock periods are allocated for the first access, and the remaining accesses are performed each in a single cycle. The first function the interface must perform is to generate the toggled A2 and A3 addresses from the first address put out by the MC68040. This involves creating a modulo 4 counter where the addresses will increment and wrap around. The MC68040 uses the burst access to fetch four long words for the internal cache line. It will start anywhere in the line so that the first data that is accessed can be passed to the processor while the rest of the data is fetched in parallel. This improves performance by fetching the immediate data first, but it does complicate the address generation logic — a standard 2 bit counter is not applicable. A typical circuit is shown.

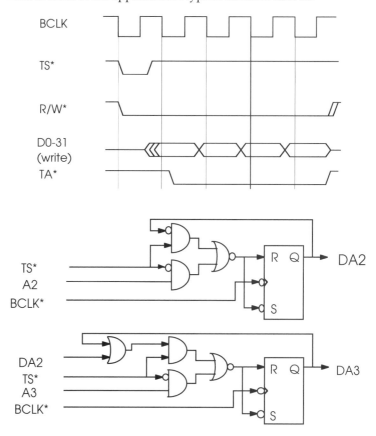

Modulo 4 counter (based on a design by John Hansen, Motorola Austin)

Given the generated addresses, the hardest task for the interface is to create the write pulse needed to write data to the FSRAMs.
The first hurdle is to ensure that the write pulse commences after the
addresses have been generated. The easiest way of doing this is to use
the two phases of the BCLK* to divide the timing into two halves.
During the first part, the address is latched by the rising edge of
BCLK*.

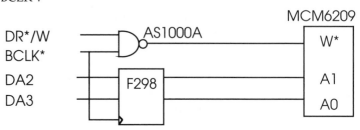

*Latching the address and gating W**

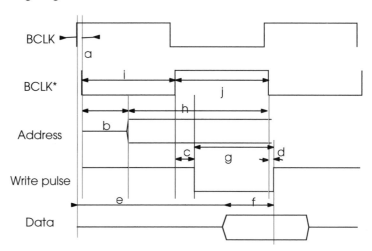

Timing	Description
a	Clock skew between BCLK and its inverted signal BCLK*.
b	Delay between BCLK* and valid address — determined by latch delay.
c	Gate delay in generating Write pulse from rising BCLK* edge.
d	Gate delay in terminating Write pulse from falling BCLK* edge.
e	Time from rising edge of BCLK to valid data from MC68040.
f	Data set-up time for write referenced from = i+j-e+a
g	Write pulse width = j-c+d.
h	Valid address, i.e. memory access time.
i,j	Cycle times for BCLK and BCLK*.

Write pulse tmings

Latching DA2 and DA3 holds the address valid while allowing the modulo 4 counter to propagate the next value through. The falling edge of BCL* is then used to gate the read/write signal to create a write pulse. The write pulse is removed before the next address is latched. This guarantees that the write pulse will be generated after the address has become valid. This circuit neatly solves the competing criteria of bringing the write pulse high before the address can be changed and the need to change the address as early as possible

The table shows the timing and the values for the write pulse, t_{WLWH}, write data set-up time, t_{DVWH} and the overall access time t_{AVAV}. For both 25 and 33 MHz speeds, the access time is always greater than 20 ns and therefore 20 ns FSRAM would be sufficient. The difficulty comes in meeting the write pulse and data set-up times. At 25 MHz, the maximum write pulse is 17 ns and the data set-up is 9 ns. Many 20 ns FSRAMs specify the minimum write pulse width with the same value as the overall access time. As a result 20 ns access time parts would not meet this specification. The data set up is also longer and it is likely that 15 ns or faster parts would have to be used. At 33 MHz, the problem is worse.

4   Basic peripherals

This chapter describes the basic peripherals that most microcontrollers provide. It covers parallel ports which are the simplest I/O devices, timer counters for generating and measuring time- and count -based events, serial interfaces and DMA controllers.

Parallel ports

Parallel ports provide the ability to input or output binary data with a single bit allocated to each pin within the port. They are called parallel ports because the initial chips that provided this support grouped several pins together to create a controllable data port similar to that used for data and address buses. It transfers multiple bits of information simultaneously, hence the name parallel port. Although the name implies that the pins are grouped together, the individual bits and pins within the port can usually be used independently of each other.

These ports are used to provide parallel interfaces such as the Centronics printer interface, output signals to LEDs and alpha-numeric displays and so on. As inputs, they can be used with switches and keyboards to support control panels.

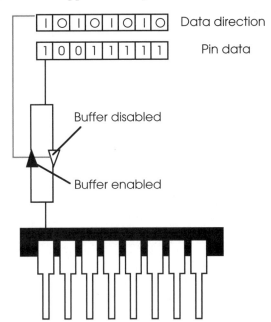

A simple parallel I/O port

The basic operation is shown in the diagram which depicts an 8 pin port. The port is controlled by two registers: a data direction register which defines whether each pin is an output or an input and a data register which is used to set an output value by writing to it and to obtain an input value by reading from it. The actual implementation typically uses a couple of buffers which are enabled depending on the setting of the corresponding bit in the data direction register.

This simple model is the basis of many early parallel interface chips such as the Intel 8255 and the Motorola 6821 devices. The model has progressed with the incorporation of a third register or an individual control bit that provides a third option of making the pin become high impedance and thus neither an input or output. This can be implemented by switching off both buffers and putting their connections to the pin in a high impedance state. Output ports that can do this are often referred to as tri-state because they can either be logic high, logic low or high impedance. In practice, this is implemented on chip as a single buffer with several control signals from the appropriate bits within the control registers. This ability has led to the development of general-purpose ports which can have additional functionality to that of a simple binary input/output pin.

Multi-function I/O ports

With many parallel I/O devices that are available today, either as part of the on-chip peripheral set or as an external device, the pins are described as general-purpose and can be shared with other peripherals. For example, a pin may be used as part of a serial port as a control signal.

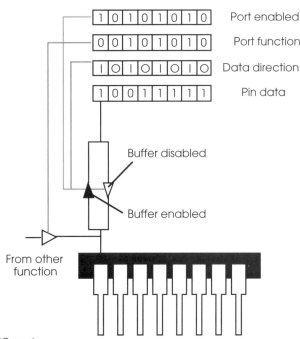

A general-purpose parallel I/O port

It may be used as a chip select for the memory design or simply as an I/O pin. The function that the pin performs is set up internally through the use of a function register which internally configures how the external pin is connected internally. If this is not set up correctly, then despite the correct programming of the other registers, the pin will not function as expected.

Note: This shared use does pose a problem for designers in that many manufacturer data sheets will specify the total number of I/O pins that are available. In practice, this is often reduced because pins need to be assigned as chip selects and to other essential functions. As a result, the number that is available for use as I/O pins is greatly reduced.

Pull-up resistors

It is important to check if a parallel I/O port or pin expects an external pull-up resistor. Some devices incorporate it internally and therefore do not need it. If it is needed and not supplied, it can cause incorrect data on reading the port and prevent the port from turning off an external device.

Timer/counters

Digital timer/counters are used throughout embedded designs to provide a series of time or count related events within the system with the minimum of processor and software overhead. Most embedded systems have a time component within them such as timing references for control sequences, to provide system ticks for operating systems and even the generation of waveforms for serial port baud rate generation and audible tones.

They are available in several different types but are essentially based around a simple structure as shown.

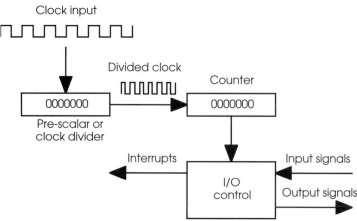

Generic timer/counter

The central timing is derived from a clock input. This clock may be internal to the timer/counter or be external and thus connected via a separate pin. The clock may be divided using a simple divider which can provide limited division normally based on a power of two or through a pre-scalar which effectively scales down or divides the clock by the value that is written into the pre-scalar register. The divided clock is then passed to a counter which is normally configured in count down operation, i.e. it is loaded with a preset value which is clocked down towards zero. When a zero count is reached, this cause an event to occur such as an interrupt of an external line changing state. The final block is loosely described as an I/O control block but can be more sophisticated than that. It generates interrupts and can control the counter based on external signals which can gate the countdown and provide additional control. This expands the functionality that the timer can provide as will be explained later.

Types

Timer/counters are normally defined in terms of the counter size that they can provide. They typically come in 8, 16 and 24 bit variants. The bit size determines two fundamental properties:

- The pre-scalar value and hence the frequency of the slowest clock that can be generated from a given clock input.

- The counter size determines the maximum value of the counter-derived period and when used with an external clock, the maximum resolution or measurement of a time-based event.

These two properties often determine the suitability of a devive for an application.

8253 timer modes

A good example of a simple timer is the Intel 8253 which is used in the IBM PC. The device has three timer/counters which provide a periodic 'tick' for the system clock, a regular interrupt every 15 μs to perform a dynamic memory refresh cycle and, finally, a source of square waveforms for use as audio tones with the built-in speaker. Each timer/counter supports six modes which cover most of the simple applications for timer/counters.

Interrupt on terminal count

This is known as mode 0 for the 8253 and is probably the simplest of its operations to understand. An initial value is loaded into the counter register and this then immediately starts to count down at the frequency determined by the clock input. When the counter reaches zero, an interrupt is generated.

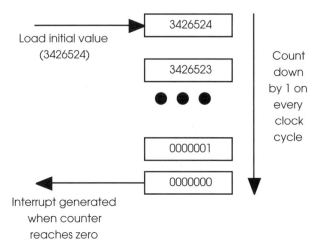

Interrupt on terminal count operation

Programmable one-shot

With mode 1, it is possible to create a single pulse with a programmable duration. The pulse length is first loaded into the counter. Nothing further happens until the external gate signal is pulled high. This rising edge starts the counter to count down towards zero and the counter output signal goes high to start the external pulse. When the counter reaches zero, the counter output goes low thus ending the pulse.

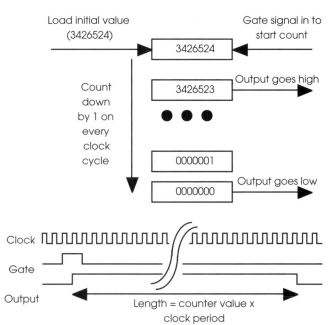

Programmable one-shot timer counter mode

The pulse duration is determined by the initial value loaded into the counter times the clock period. While this is a common timer/counter mode, many devices such as the 8253 incorporate a reset. If the gate signal is pulled low and then high again to create a new rising edge while the counter is counting down, the current count value is ignored and replaced by the initial value and the count continued. This means that the output pulse length will be extended by the time between the two gate rising edges.

This mode can be used to provide pulse width modulation for power control where the gate is connected to a zero crossing or similar detector or clock source to create a periodic signal.

Rate generator

This is a simple divide by N mode where N is defined by the initial value loaded into the counter. The output from the counter consists of a single low with the time period of a single clock followed by a high period controlled by the counter which starts to count down. When the counter reaches zero, the output is pulled low, the counter reloaded with the initial value and the process repeated. This is mode 3 with the 8253.

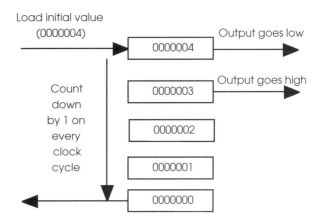

Rate generation (divide by N)

Square wave rate generator

Mode 4 is similar to mode 3 except that the waveform is a square wave with a 50:50 mark/space ratio. This is achieved by extending the low period and by reducing the high period to half the clock cycles specified by the initial counter value.

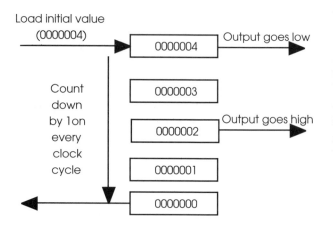

Square wave generation

Software triggered strobe

When mode 4 is enabled, the counter will start to count as soon as it is loaded with its initial value. When it reaches zero, the output is pulsed low for a single clock period and then goes high again. If the counter is reloaded by software before it reaches zero, the output does not go low. This can be used as a software-based watchdog timer where the output is connected to a non-maskable interrupt line or a system reset.

Hardware triggered strobe

Mode 5 is similar to mode 4 except that the retriggering is done by the external gate pin acting as a trigger signal.

Generating interrupts

The 8253 has no specific interrupt pins and therefore the timer OUT pin is often used to generate an external interrupt signal. With the IBM PC, this is done by connecting the OUT signal from timer/counter 0 to the IRQ 0 signal and setting the timer/counter to run in mode 3 to generate a square wave. The input clock is 1.19318 MHz and by using a full 16 bit count value, is divided by 65536 to provide a 18.3 Hz timer tick. This is counted by the software to provide a time of day reference and to provide a system tick.

MC68230 modes

The Motorola MC68230 is a good example of a more powerful timer architecture that can provide a far higher resolution than the Intel 8253. The timer is based around a 24 bit architecture which is split into three 8 bit components. The reason for this is that the device uses an 8 bit bus to communicate with the host processor such as a MC68000 CPU. This means that the counter cannot be loaded directly from the processor in a single bus cycle. As a result, three preload registers have been added to the basic architecture previously described. These are preloaded using three separate accesses prior to writing to the Z control bit in the control register. This transfers the contents of the preload register to the counter as a single operation.

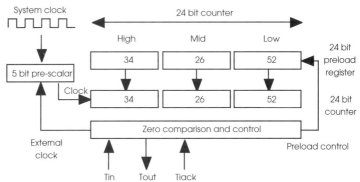

The MC68230 timer/counter architecture

Instead of writing to the counter to either reset it or initialise it, the host processor uses a combination of preload registers and the Z bit to control the timer. The timer can be made to preload when it reaches zero or, as an option, simply carry on counting. This gives a bit more flexibility in that timing can be performed after the zero count as well as before it.

This architecture also has a 5 bit pre-scalar which is used to divide the incoming clock which can be sourced from the system clock or externally via the Tin signal. The pre-scalar can be loaded with any 5 bit value to divide the clock before it drives the counter.

Timer processors

An alternative to using a timer/counter is the development of timer computers where a processor is used exclusively to manage and implement complex timing functions over multiple timer channels. The MC68332 is a good example of such a processor. It has a CPU32 processor (MC68020 based) running at 16 MHz and a timer processor unit instead of a communications processor. This has 16 channels which are controlled by a RISC-like processor to perform virtually any timing function. The timing resolution is down to 250 nanoseconds with an external clock source or 500 nanoseconds with an internal one. The timer processor can perform the common timer algorithms on any of the 16 channels without placing any overhead on the CPU32.

A queued serial channel and 2 kbits of power-down static RAM are also on-chip and for many applications all that is required to complete a working system is an external program EPROM and a clock.

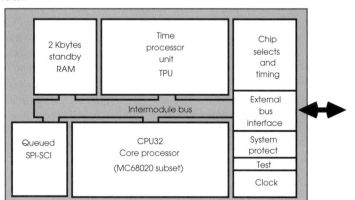

The MC68332 block diagram

The timer processor has several high level functions which can easily be accessed by the main processor by programming a parameter block. For example, the missing tooth calculation for generating ignition timing can be easily performed through a combination of the timer processor and the CPU32 core. A set of parameters is calculated by the CPU32 and loaded into a parameter block which commands the timer processor to perform the algorithm. Again, no interrupt routines or periodic peripheral bit manipulation is needed by the CPU32.

$00	REF TIME	**CHANNEL_CONTROL**
$01		
$02	**MAX_MISSING**	**NUM_OF_TEETH**
$03	BANK_SIGNAL/MISSING_COUNT	ROLLOVER_COUNT
$04		
$05	**RATIO**	**TCR2_MAX_VALUE**
	PERIOD_HIGH_WORD	
	PERIOD_LOW_WORD	

ERROR	TCR2_VALUE

☐ updated by CPU32 host

The parameter block for a period measurement
with missing transition detection

Real-time clocks

There is a special category of timer known as a real-time clock whose function is to provide an independent time keeper that can provide time measurements in terms of the current time and date as opposed to a counter value. The most popular device is probably the MC146818 and its derivatives and clones that were used in the first IBM PC. These devices are normally driven off a 32 kHz watch crystal and are battery backed up to maintain the data and time. The battery back-up was done externally with a battery or large capacitor but has also been incorporated into the chip in the case of the versions supplied by Dallas Semiconductor. These devices can also provide a system tick signal for use by the operating system.

Simulating a real-time clock in software

These can be simulated in software by programming a timer to generate a periodic event and simply using this as a reference and counting the ticks. The clock functions are then created as part of the software. When enough ticks have been received it updates the seconds counter and so on. There are two problems with this: the first concerns the need to reset the clock when the system is turned off and the second concerns the accuracy which can be quite bad. This approach does save on a special clock chip and is used on VCRs, ovens and many other appliances. This also explains why they need resetting when there has been a power cut!

Serial ports

Serial ports are a pin efficient method of communicating between other devices within an embedded system. With microcontrollers which do not have an external extension bus, they can provide the only method of adding additional functionality.

The simplest serial ports are essentially a pair of shift registers that are connected together with one input (receiver) connected to the output of the other to create a transmitter. They are clocked together by a common clock and thus data is transmitted from one register to the other. The time taken is dependent on the clock frequency and the number of bits that are transferred. The shift registers are normally 8 bits wide.

When the transmitter is emptied, it can be made to generate a local interrupt which can indicate to the processor that the byte has been transferred and/or that the next byte should be loaded into the register. The receiver can also generate an interrupt when the complete byte is received to indicate that it is ready for reading. Most serial ports use a FIFO as a buffer so that data is not lost. This can happen if data is transmitted before the preceding byte has been read. With the FIFO buffer, the received byte is transferred to it when the byte is received. This frees up the shift register to receive more bits without losing the data. The FIFO buffer is read to receive the data hence the acronym's derivation — first in, first out. The reverse can be done for the transmitter so that data can be sent to the transmitter before the previous value has been sent.

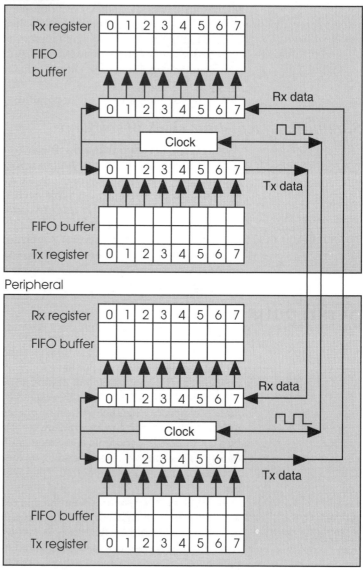

Serial interface with FIFO buffering

The size of the FIFOs is important in reducing processor overhead and increasing the serial port's throughput as will be explained in more detail later on.

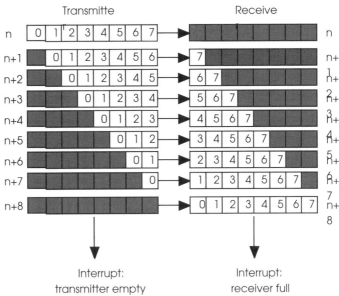

Basic serial port operation

The diagram shows a generic implementation of a serial interface between a processor and peripheral. It uses a single clock signal which is used to clock the shift registers in each transmitter and receiver. The shift registers each have a small FIFO for buffering. The clock signal is shown as being bi-directional: in practice it can be supplied by one of the devices or by the device that is transmitting. Obviously care has to be taken to prevent the clock from being generated by both sides and this mistake is either prevented by software protocol or through the specification of the interface.

Serial peripheral interface

This bus is often referred to as SPI and is frequently used on Motorola processors such as the MC68HC05 and MC68HC11 microcontrollers to provide a simple serial interface. It uses the basic interface as described in the previous section with a shift register in the master and slave devices driven by a common clock. It allows full-duplex synchronous communication between the MCU and other slave devices such as peripherals and other processors.

Data is written to the SPDR register in the master device and clocked out into the slave device SPDR using the common clock signal SCK. When 8 bits have been transferred, an interrupt is locally generated so that the data can be read before the next byte is clocked through. The SS or slave select signal is used to select which slave is to receive the data. In the first example, shown with only one slave, this is permanently asserted by grounding the signal. With multiple slaves, spare parallel I/O pins are used to select the slave prior to data transmission. The diagram shows such a configuration. If pin 1 on the master MCU is driven low, slave 1 is selected and so on. The unselected slaves tri-state the SPI connections and do not receive the clocked data and take no part in the transfer.

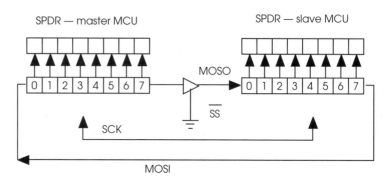

SPI internal architecture

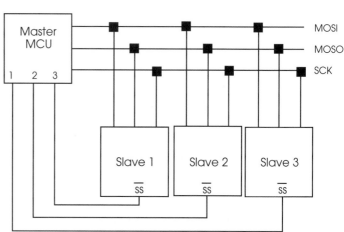

Supporting multiple slave devices

It should not be assumed that an implementation buffers data. As soon as the master writes the data into the SPDR it is transmitted as there is no buffering. As soon as the byte has been clocked out, an interrupt is generated indicating that the byte has been transferred. In addition, the SPIF flag in the status register (SPSR) is set. This flag must be cleared by the ISR before transmitting the next byte.

The slave device does have some buffering and the data is transferred to the SPDR when a complete byte is transferred. Again, an interrupt is generated when a byte is received. It is essential that the interrupt that is generated by the full shift register is serviced quickly to transfer the data before the next byte is transmitted and transferred to the SPDR. This means that there is an eight clock time period for the slave to receive the interrupt and transfer the data. This effectively determines the maximum data rate.

I²C bus

The inter-IC, or I²C bus as it is more readily known, was developed by Philips originally for use within television sets in the mid-1980s. It is probably the most known simple serial interface currently used. It combines both hardware and software protocols to provide a bus interface that can talk to many peripheral devices and can even support multiple bus masters. The serial bus itself only uses two pins for its implementation.

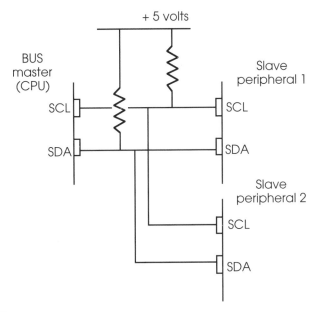

I²C electrical connections

The bus consists of two lines called SDA and SCL. Both bus masters and slave peripheral devices simply attach to these two lines as shown in the diagram. For small numbers of devices and where the distance between them is small, this connection can be direct. For larger numbers of devices and/or where the track length is large, Philips can provide a special buffer chip (P82B715) to increase the current drive. The number of devices is effectively determined by the line length, clock frequency and load capacitance which must not exceed 400 pF although derating this to 200 pF is recommended. With low frequencies, connections of several metres can be achieved without resorting to special drivers or buffers.

The drivers for the signals are bi-directional and require pull-up resistors. When driven they connect the line to ground to create a low state. When the drive is removed, the output will be pulled up to a high voltage to create a high signal. Without the pull-up resistor, the line would float and can cause indeterminate values and thus cause errors.

The SCL pin provides the reference clock for the transfer of data but it is not a free running clock as used by many other serial ports. Instead it is clocked by a combination of the master and slave device and thus the line provides not only the clock but also a hardware handshake line.

The SDA pin ensures the serial data is clocked out using the SCL line status. Data is assumed to be stable on the SDA line if SCL is high and therefore any changes occur when the SCL is low. The sequence and logic changes define the three messages used.

Message	1st event	2nd event
START	SDA HÆL	SCL HÆL
STOP	SCL LÆH	SDA LÆH
ACK	SDA HÆL	SCL HÆL

The table shows the hardware signalling that is used for the three signals, START, STOP and ACKNOWLEDGE. The START and ACKNOWLEDGE signals are similar but there is a slight difference in that the START signal is performed entirely by the master whereas the ACKNOWLEDGE signal is a handshake between the slave and master.

Data is transferred in packets with a packet containing one or more bytes. Within each byte, the most significant bit is transmitted first. A packet, or telegram as it is sometimes referred to, is defined as the data transmitted between START and STOP signals sent from the master. Within the packet transmission, the slave will acknowledge each byte by using the ACKNOWLEDGE signal. The basic protocol is shown in the diagram.

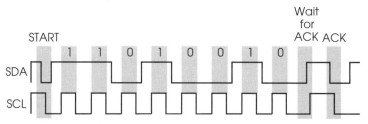

Write byte transfer with ACKNOWLEDGE

The 'wait for ACK' stage looks like another data bit except that it is located as the ninth bit. With all data being transmitted as bytes, this extra one bit is interpreted by the peripheral as an indication that the slave should acknowledge the byte transfer. This is done by the slave pulling the SDA line low after the master has released the data and clock line. The ACK signal is physically the same as the START signal and is ignored by the other peripherals because the data packet has not been terminated by the STOP command. When the AC-KNOWLEDGE command is issued, it indicates that the transfer has been completed. The next byte of data can start transmission by pulling the SCL signal down low.

Read and write access

While the previous paragraphs described the general method of transferring data, there are some specific differences for read and write accesses. The main differences are concerned with who controls which line during the handshake.

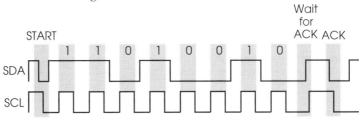

Write byte transfer with ACKNOWLEDGE

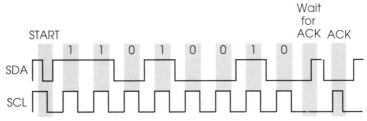

Read byte transfer with ACKNOWLEDGE

During a write access the sequence is as follows:

- After the START and 8 bits have been transmitted, the master releases the data line followed by the clock line. At this point it is waiting for an acknowledgement.
- The addressed slave will pull the data line down to indicate the ACKNOWLEDGE signal.
- The master will drive the clock signal low and in return, the slave will release the data line, ready for the first bit of the next byte to be transferred or to send a STOP signal.

During a read access the sequence is as follows:

- After the 8 bits have been transmitted by the slave, the slave releases the data line.
- The master will now drive the data line low.
- The master will then drive the clock line high and low to create a clock pulse.
- The master will then release the data line already for the first bit of the next byte or a STOP signal.

It is also possible to terminate a transfer using a STOP instead of waiting for an ACKNOWLEDGE. This is sometimes needed by some peripherals which do not issue an ACKNOWLEDGE on the last transfer. The STOP signal can even be used in mid transmission of the byte if necessary.

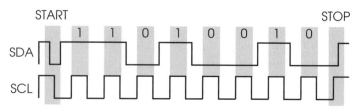

A Write byte transfer with STOP

Addressing peripherals

As mentioned before, the bus will support multiple slave devices. This immediately raises the question of how the protocol selects a peripheral. All the devices are connected onto the two signals and therefore can see all the transactions that occur. The slave selection is performed by using the first byte within the data packet as an address byte. The protocol works as shown in the diagram. The master puts out the START signal and this tells all the connected slave devices to start accepting the data. The address byte is sent out and each slave device compares the address with its own value. If there is a match, then it will send the ACKNOWLEDGE signal. If there is no match, then there has been a programming error. In this case, there will be no ACKNOWLEDGE signal returned and effectively the SDA signal will remain high.

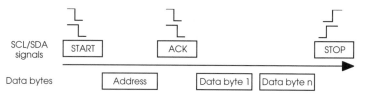

A complete data packet including addressing and signalling

The address value is used to select the device and to indicate the type of operation that the master requests: read if the eighth bit is set to one or write if set to zero. This means that the address byte allows 128 devices with read/write capability to be connected. The address values for a device is either pre-programmed, i.e. assigned to that device number, or can be programmed through the use of external pins. Care must be made to ensure that no two devices have the same address.

In practice, the number of available addresses is less because some are reserved and others are used as special commands. To provide more addressing, an extended address has been developed that uses two bytes: the first byte uses a special code (five ones) to distinguish it from a single byte address. In this way both single byte and double byte address slaves can be used on the same bus.

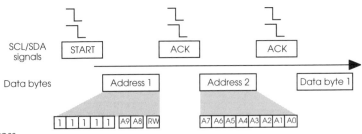

Sending a 2 byte address

Sending an address index

So far the transfers that have been described have assumed that the peripheral only has one register or memory location. This is really the case and thus several addressing schemes have been developed to address individual locations within the peripheral itself.

For peripherals with a small number of locations, a simple technique is simply to incorporate an auto-incrementing counter within the peripheral so that each access selects the next register. As the diagram shows, the contents of register 4 can be accessed by performing four successive transfers after the initial address has been sent.

Auto-incrementing access

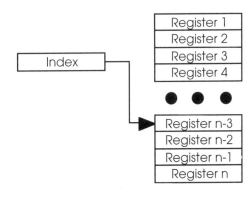

Combined format

This is fine for small numbers of registers but with memory devices such as EEPROM this is not an efficient method of operation. An alternative method uses an index value which is written to the chip, prior to accessing the data. This is known as the combined format and uses two data transfers. The first transfer is a write with the index value that is to be used to select the location for the next access. To access memory byte 237, the data byte after the address would contain 237. The next transfer would then send the data in the case of a write or request the contents in the case of a read. Some devices also support auto-incrementing and thus the second transfer can access multiple sequential locations starting at the previously transmitted index value.

Timing

One of the more confusing points about the bus is the timing or lack of it. The clock is not very specific about its timings and does not need a specified frequency or even mark to space ratios. It can even be stopped and restarted at a later point in time if needed. The START, STOP and ACKNOWLEDGE signals have a minimum delay time between the clock and data edges and pulse widths but apart from this, everything is very free and easy.

This is great in one respect but can cause problems in an other. Typically the problem is concerned with waiting for the ACKNOWL-EDGE signal before proceeding. If this signal is not returned then the bus will be locked up until the master terminates the transfers with a STOP signal. It is important therefore not to miss the transition. Unfortunately, the time taken for a slave to respond is dependent on the peripheral and with devices like EEPROM, especially during a write cycle, this time can be extremely long.

As a result, the master should use a timer/counter to determine when sufficient time has been given to detect the ACKNOWL-EDGE before issuing a STOP.

This can be done in several ways: polling can be used with a counter to determine the timeout value. The polling loop is completed when either the ACKNOWLEDGE is detected to give a success or if the polling count is exceeded. An alternative method is to use a timer to tell the master when to check for the acknowledgement. There are refinements that can be added where the timeout values are changed depending on the peripheral that is being accessed. A very sophisti-cated system can use a combination of timer and polling to check for the signal n times with an interval determined by the timer. Which-ever method is chosen, it is important that at least one is implemented to ensure that the bus is not locked up.

Multi-master support

The bus supports the use of multiple masters but does not have any in-built mechanism for controlling access. Instead, it uses a technique called collision detect to determine if two masters start to use the bus at the same time. A master wait until the bus is clear, i.e. there is no current transfer, and then issue a START signal. If another master has done the same then the signals that appear on the line will be corrupted. If a master wants a line to be high and the other wants to drive it low, then the line will go low. With the bi-directional ports that are used, each master can monitor the line and confirm that it is in the expected state. If it is not, then a collision has occurred and the master should discontinue transmission and allow the other master to continue.

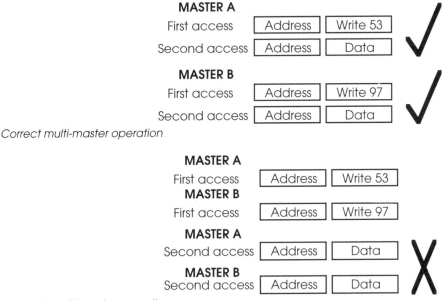

Correct multi-master operation

Incorrect multi-master operation

It is important that timeouts for acknowledgement are incorporated to ensure that the bus cannot be locked up. In addition, care must be taken with combined format accesses to prevent a second master from resetting the index on the peripheral. If master A sets the index into an EEPROM peripheral to 53 and before it starts the next START-address-data transfer, a second master gets the bus and sets the index to its value of 97, the first master will access incorrect data. The problem can be even worse as the diagram shows. When master B overwrites the index value prior to master A's second access, it causes data corruption for both parties. Master A will access location 97 and due to auto-incementing, master B will access location 98 — neither of which is correct! The bus does not provide a method of solving this dilemma and the only real solutions are not to share the peripheral between the devices or use a semaphore to protect access. The protection of resources is a perennial problem with embedded systems and will be covered in more detail later on.

M-Bus (Motorola)

M-Bus is an ideal interface for EEPROMs, LCD controllers, A/D converters and other components that could benefit from fast serial transfers. This two-wire bi-directional serial bus allows a master and a slave to rapidly exchange data. It allows for fast communication with no address translation. It is very similar in operation to I²C and thus M-Bus devices can be used with these type of serial ports. The maximum transfer rate is 100 kb/s.

What is an RS232 serial port?

Up until now, the serial interfaces that have been described have used a clock signal as a reference and therefore the data transfers are synchronous to that clock. For the small distances between chips, this is fine and the TTL or CMOS logic voltages are sufficient to ensure operation over the small connection distances. However, this is not the case if the serial data is being transmitted over many metres. The low voltage logic levels can be affected by the cable capacitance and thus a logic one at the transmitter may be seen as an indeterminate voltage at the receiver end. Clock edges can become skewed and out of sync with the data causing the wrong data to be accepted. As a result, a slightly different serial port is used for connecting over longer distance, generically referred to a RS232.

For most people, the mention of RS232 immediately brings up the image and experiences of connecting peripherals to the ubiquitous IBM PC. The IBM PC typically has one or two serial ports, *COM1* and *COM2*, which are used to transfer data between the PC and printers, modems and even other computers. The term 'serial' comes from the fact that only one data line is used to transmit and receive data and thus the information must be sent and received a bit at a time. Instead of transmitting the 8 bits that make up a byte using eight data lines at once, one data line is used to send 8 bits, one at a time. In practice, several lines are used to provide separate lines for data transmit and receive, and to provide a control line for hardware handshaking. One important difference is that the data is transmitted asynchronously i.e. there is no separate reference clock. Instead the data itself provides its own reference clock in terms of its format.

The serial interface can be divided into two areas. The first is the physical interface, commonly referred to as RS232 or EIA232, which is used to transfer data between the terminal and the computer. The electrical interface uses a combination of +5, +12 and –12 volts for the electrical interface. This used to require the provision of additional power connections but there are now available interface chips that take a 5 volt supply (MC1489) and generate internally the other voltages that are needed to meet the interface specification. Typically, a logic one is signalled by a +3 to +15 volts level and a logic zero by –3 to –15 volts. Many systems use +12 and –12 volts.

Note: The term RS232 strictly specifies the physical interface and not the serial protocol. Partly because RS232 is easier to say than universal asynchronous communication using an RS232 interface, the term has become a general reference to almost any asynchronous serial communication.

The second area controls the flow of information between the terminal and computer so that neither is swamped with data it cannot handle. Again, failure to get this right can cause data corruption and other problems.

When a user presses a key, quite a lengthy procedure is carried out before the character is transmitted. The pressed key generates a specific code which represents the letter or other character. This is converted to a bit pattern for transmission down the serial line via the serial port on the computer system. The converted bit pattern may contain a number of start bits, a number of bits (5, 6, 7 or 8) representing the data, a parity bit for error checking and a number of stop bits. These are all sent down the serial line by a UART (universal asynchronous receiver transmitter) in the terminal at a predetermined speed or baud rate.

Serial bit stream

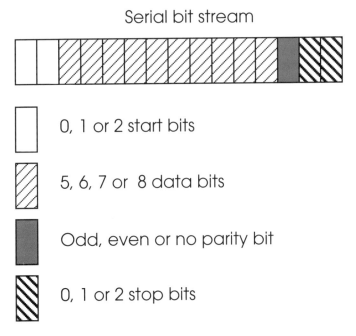

Serial bit streams

The start bits are used to indicate that the data being transmitted is the start of a character. The stop bits indicate that character has ended and thus define the data sequence that contains the data. The parity bit can either be disabled, i.e. set to zero or configured to support odd or even parity. The bit is set to indicate that the total number of bits that have been sent is either an odd or even number.

This allows the receiving UART to detect a single bit error during transmission or reception. The bit sequencing and resultant waveform is asynchronous in that there is not a reference clock transmitted. The data is detected by using a local clock reference, i.e. from the baud rate generator and the start/stop bit edges. This is why it is so important not only to configure the data settings but to set the correct baud rate settings so that the individual bits are correctly interpreted. As a result, both the processor and the peripheral it is communicating with must use the same baud rate and the same combination of start, stop, data and parity bits to ensure correct communication. If different combinations are used, data will be wrongly interpreted.

If the terminal UART is configured in half duplex mode, it echoes the transmitted character so it can be seen on the screen. Once the data is received at the other end, it is read in by another UART and, if this UART is set up to echo the character, it sends it back to the terminal. (If both UARTs are set up to echo, multiple characters are transmitted!) The character is then passed to the application software or operating system for further processing.

If the other peripheral or processor is remote, the serial line may include a modem link where the terminal is connected to a modem and a telephone line, and a second modem is linked to the computer at the other end. The modem is frequently controlled by the serial line, so if the terminal is switched off, the modem effectively hangs up and disconnects the telephone line. Modems can also echo characters and it is possible to get four characters on the terminal screen in response to a single key stroke.

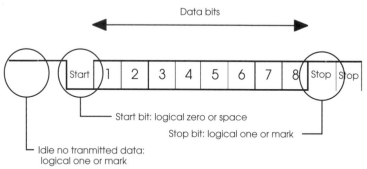

Asynchronous data format

The actual data format for the sequence is shown in the diagram. When no data is transmitted, the TXD signal is set to a logical one. When data is transmitted, a start bit is sent by setting the line to a logical zero. Data is then sent by setting the data to a zero or one accordingly and finally the stop bits are sent by forcing the line to a logical one. The stop bits essentially look the same as the idle bits when no data is being transmitted. The timing is defined by the baud rate that both the receiver and transmitter are using. The baud rate used to be supplied by an external timer/counter called a baud rate generator that generates a clock signal at the right frequency. This function is now performed on-chip with modern controller chips and usually can work with the system clock or with a simple watch crystal instead of one with a specific frequency.

Note: If the settings are slightly incorrect, i.e. the number of stop and data bits iswrong, then it is possible for the data to appear to be received correctly. For example, if data is transmitted at 7 data bits with 2 stop bits and received as 8 data bits with 1 stop bit, the receiver would get the 7 data bits and set the eighth data bit to a one. If this character was then displayed on the screen, it could appear in the correct format due to the fact that many character sets ignore the eighth bit. In this case, the software and system would appear to work. If the data was used in some other protocol where the eighth bit has either used or assumed to be set to zero, the program and system would fail!

Asynchronous flow control

Flow control is necessary to prevent either the terminal or the computer from sending more data than the other can cope with. If too much is sent, it either results in missing characters or in a data overrun error message. The first flow control method is hardware handshaking, where hardware in the UART detects a potential overrun and asserts a handshake line to tell the other UART to stop transmitting. When the receiving device can take more data, the handshake line is released. The problem with this is that there are several options and, unless the lines are correctly connected, the handshaking does not work correctly and data loss is possible. The second method uses software to send flow control characters XON and XOFF. XOFF stops a data transfer and XON restarts it. Unfortunately, there are many different ways of using these lines and, as a result, this so-called standard has many different implementations.

The two most common connectors are the 25 pin D type and the 9 pin D type. These are often referred to as DB-25 and DB-9 respectively. Their pin assignments are as follows:

DB-25	Signal	DB-9
1	Chassis ground	Not used
2	Transmit data — *TXD*	3
3	Receive data — *RXD*	2
4	Request to send — *RTS*	7
5	Clear to send — *CTS*	8
6	Data set ready — *DSR*	6
7	Signal ground — *GND*	5
8	Data carrier detect — *DCD*	1
20	Data terminal ready — *DTR*	4
22	Ring indicator — *RI* or *RING*	9

There are many different methods of connecting these pins and this has caused many problems especially for those faced with the task of implementing the software for a UART in such a configuration. To implement hardware handshaking, individual I/O pins are used to act as inputs or outputs for the required signals. The functionality of the various signals is as follows:

TXD   Transmit data. This transmits data and would normally be connected to the RXD signal on the other side of the connection.

RXD   Receive data. This transmits data and would normally be connected to the TXD signal on the other side of the connection. In this way, there is a cross-over connection.

RTS   Request to send. This is used in conjunction with CTS to indicate that this side is ready to send and needs confirmation that the other side is ready.

CTS   Clear to send. This is the corresponding signal to RTS and is sent by the other side on receipt of the RTS to indicate that it is ready to receive data.

DSR   Data set ready. This is used in conjunction with DTR to indicate that each side is powered on and ready.

DCD   Data carrier detect. This is normally used to determine which side is in control of the hardware handshake protocol.

DTR   Data terminal ready. This is used in conjunction with DSR to indicate that each side is powered on and ready.

RI   Ring indicator.This is asserted when a connected modem has detected an incoming call.

Much of the functionality of these signals has been determined by the need to connect to modems initially to allow remote communication across telephone lines. While modem links are still important, many serial lines are used in modemless links to peripherals such as printers. In these cases, the interchange of signals which the modem performs must be simulated within the cabling and this is done using a null modem cable. The differences are best shown by looking at some example serial port cables.

Modem cables

These are known as modem or straight through cables because the connections are simply one to one with no crossing over or other more complex wiring. They are used to link PCs with modems, printers, plotters and other peripherals. However, do not use them when linking a PC to another PC or computer — they won't work! For those links, a null modem cable is needed.

Null modem cables

Null modem cables are used to link PCs together. They work by switching over the transmit and receive signals and the handshaking connections so that each PC 'sees' a modem at the other end. There are many configurations depending on the number of wires that are needed within the cable.

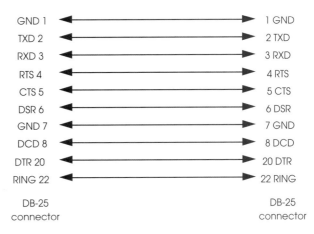

An IBM PS/2 and PC XT to modem cable

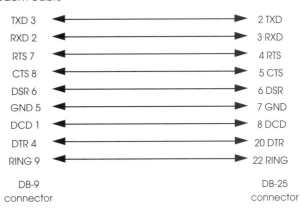

An IBM PC AT to modem cable

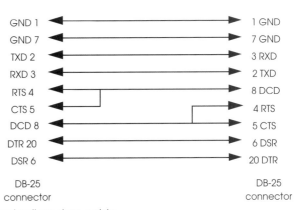

An IBM DB-25 to DB-25 standard null modem cable

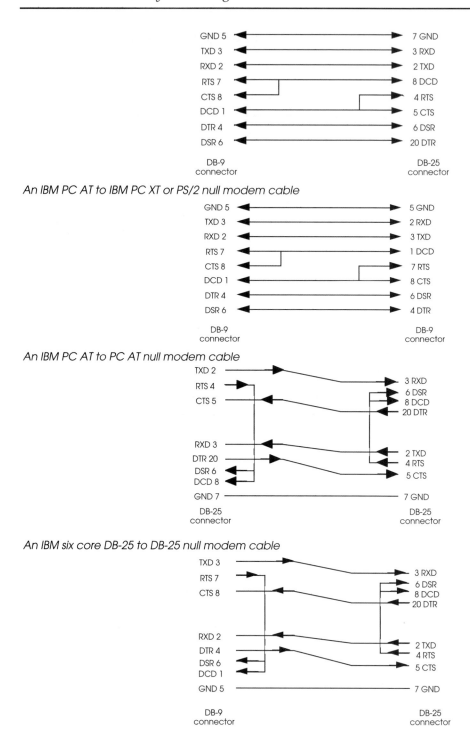

GND 5 ◄─────────────────► 7 GND
TXD 3 ◄─────────────────► 3 RXD
RXD 2 ◄─────────────────► 2 TXD
RTS 7 ◄─────────────────► 8 DCD
CTS 8 ◄─────────────────► 4 RTS
DCD 1 ◄─────────────────► 5 CTS
DTR 4 ◄─────────────────► 6 DSR
DSR 6 ◄─────────────────► 20 DTR

DB-9
connector

DB-25
connector

An IBM PC AT to IBM PC XT or PS/2 null modem cable

GND 5 ◄─────────────────► 5 GND
TXD 3 ◄─────────────────► 2 RXD
RXD 2 ◄─────────────────► 3 TXD
RTS 7 ◄─────────────────► 1 DCD
CTS 8 ◄─────────────────► 7 RTS
DCD 1 ◄─────────────────► 8 CTS
DTR 4 ◄─────────────────► 6 DSR
DSR 6 ◄─────────────────► 4 DTR

DB-9
connector

DB-9
connector

An IBM PC AT to PC AT null modem cable

TXD 2 ──────────────────► 3 RXD
RTS 4   6 DSR
CTS 5   8 DCD
                          20 DTR

RXD 3   2 TXD
DTR 20   4 RTS
DSR 6   5 CTS
DCD 8
GND 7 ────────────────── 7 GND

DB-25
connector

DB-25
connector

An IBM six core DB-25 to DB-25 null modem cable

TXD 3 ──────────────────► 3 RXD
RTS 7   6 DSR
CTS 8   8 DCD
                          20 DTR

RXD 2   2 TXD
DTR 4   4 RTS
DSR 6   5 CTS
DCD 1
GND 5 ────────────────── 7 GND

DB-9
connector

DB-25
connector

An IBM six core DB-9 to DB-25 null modem cable

XON-XOFF flow control

Connecting these wires together and using the correct pins is not a trivial job and two alternative approaches have been developed. The first is the development of more intelligent UARTs that handle the flow control directly with little or no intervention from the processor. The second is to dispense with hardware handshaking completely and simply use software handshaking where characters are sent to control the flow of characters between the two systems. This latter approach is used by Apple Macintosh, UNIX and many other systems because of its reduced complexity in terms of the hardware interface and wiring.

With the XON-XOFF protocol, an XOFF character (control-S or ASCII code 13) is sent to the other side to tell it to stop sending any more data. The halted side will wait until it receives an XON character (control-Q or ASCII code 11) before recommencing the data transmission. This technique does assume that each side can receive the XOFF character and that transmission can be stopped before overflowing the other side's buffer.

UART implementations

8250/16450/16550

Probably the most commonly known and used UART is the 8250 and its derivatives as used to provide serial ports COM1 and COM2 on an IBM PC. The original design used an Intel 8250 which has been largely replaced by the National Semiconductor 16450 and 16550 devices or by cloned devices within a Super I/O chip which combines all the PC's I/O devices into a single piece of silicon.

D0	1		40	VCC
D1	2		39	*RI
D2	3		38	*DCD
D3	4		37	*DSR
D4	5		36	*CTS
D5	6		35	MR
D6	7		34	*OUT1
D7	8		33	*DTR
RCLK	9		32	*RTS
SIN	10	31	*OUT2	
SOUT	11	30	INTR	
CS0	12	29	NC (-RXRDY)	
CS1	13	28	A0	
*CS2	14	27	A1	
*BAUDOUT	15	26	A2	
XIN	16	25	*ADS	
XOUT	17	24	CSOUT (-TXRDY)	
*WR	18	23	DDIS	
WR	19	22	RD	
VSS	20	21	*RD	

* indicates an active low signal

UART pinout

The original devices used voltage level shifters to provide the + and –12 volt RS232 signalling voltage levels but this function is sometimes included within the UART as well.

The pinout shows the hardware signals that are used and these fall into two groups: those that are used to provide the UART interface to the processor and those that are the UART signals. Some of the signals are active low, i.e. when they are at a zero voltage level, they represent a logical one. These signals are indicated by an asterisk.

The interface signals

*ADS
: This is the address strobe signal and is used to latch the address and chip select signals during a processor access. The latching takes place on the positive edge of the and assumes that the other signals are stable at this point. This signal can be ignored by permanently asserting it. In this case, the address and chip selects must be set up and stable for the whole cycle with the processor and peripheral clock signals providing the timing references. The IBM PC uses the chip in this way.

*BAUDOUT
: This is the 16x clock signal from the transmitter section of the UART. The clock frequency is the main clock frequency divided by the values stored in the baud generator divisor latches. It is normally used — as in the IBM PC, for example — to route the transmit clock back into the receive section by connecting this pin to the RCLK pin. By doing this, both the transmit and receive baud rates are the same and use the same clock frequency. To create an asynchronous system such as 1200/75 which is used for teletext links, an external transmit clock is used to feed RCLK instead.

CS0,1 and 2
: These signals are used to select the UART and are derived from the rest of the processor's address signals. The lower 3 bits of the CPU address bus are connected to the A0–A2 pins to select the internal registers. The rest of the address bus is decoded to generate a chip select signal. This can be a single entity, in which case two of the chip selects are tied to the appropriate logic level. If the signal is low, then CS0 and CS1 would be tied high. The provision of these three chip selects provides a large amount of flexibility. The truth table is shown overleaf.

CS0	CS1	CS2	Action
High	High	Low	Selected
Low	Low	Low	Dormant
Low	Low	High	Dormant
Low	Low	Low	Dormant
Low	High	High	Dormant
Low	High	Low	Dormant
High	Low	High	Dormant
High	Low	Low	Dormant

D0–D7   These signals form the 8 bit bus that is connected between the peripheral and the processor. All transfers between the UART and processor are byte based.

DDIS   This goes low whenever the CPU is reading data from the UART. It can be used to control bus arbitration logic.

INTR   This pin is normally connected to an interrupt pin on the processor or in the case of the IBM PC, the interrupt controller. It is asserted when the UART needs data to be transferred to or from the internal buffers, or if an error condition has occurred such as a data overrun. The ISR has to investigate the UART's status registers to determine the actual service(s) requested by the peripheral.

MR   This is the master reset pin and is used to reset the device and restore the internal registers to their power-on default values. This is normally connected to the system/processor reset signal to ensure that the UART is reset when the system is.

*OUT1   This is a general-purpose I/O pin whose state can be set by programming bit 2 of the MCR to a '1'.

*OUT2   This is another general-purpose I/O pin whose state can be set by programming bit 3 of the MCR '1'. In the IBM PC it is used to gate the interrupt signal from the UART to the interrupt controller. In this way, interrupts from the UART can be externally disabled.

RCLK   This is the input for the clock for the receiver section of the chip. See •BAUDOUT on the previous page for more details.

RD, *RD			These are read strobes that are used to indicate the type of access that the CPU needs to perform. If RD is high or *RD is low, the CPU access is a read cycle.
SIN			This is the serial data input pin for the receiver.
SOUT			This is the serial data output pin for the transmitter.

*RXRDY,*TXRDY

These pins are used for additional DMA control and can be used to initiate DMA transfers to and from the read and write buffers. They are not used within the IBM PC design where the CPU is responsible for moving data to and from the UART.

WR, *WR   These are read strobes that are used to indicate the type of access that the CPU needs to perform. If WR is high or *WR is low, the CPU access is a write cycle.

XIN, XOUT   These pins are used to either connect an external crystal or connect to an external clock. The frequency is typically 8 MHz.

A0–2   These are the three address signals which are used in conjunction with DLAB to select the internal registers. They are normally connected to the lower bits of the processor address bus. The upper bits are normally decoded to create a set of chip select signals to select the UART and locate it at a specific address location.

DLAB	A2	A1	A0	Register
0	0	0	0	READ: receive buffer
				WRITE: transmitter holding
0	0	0	1	Interrupt enable
x	0	1	0	READ: Interrupt identification
				WRITE: FIFO control *
x	0	1	1	Line control
x	1	0	0	Modem control
x	1	0	1	Line status
x	1	1	0	Modem status
x	1	1	1	Scratch
1	0	0	0	Divisor latch (LSB)
1	0	0	1	Divisor latch (MSB)

*undefined with the 16450.

Register descriptions

The main difference between the various devices concerns the buffer size that they support and, in particular, the effect that it has on the effective throughput of the UART.

The UART relies on the CPU to transfer data and therefore the limit on the serial data throughput that can be sustained is determined by the time it takes to interrupt the CPU and for the appropriate interrupt service routine to identify the reason for the interrupt — it may have been raised as a result of an error — and then transfer the data if the interrupt corresponds to a data ready for transfer request. Finally, the processor returns from the interrupt.

The time to perform this task is determined by the processor type and memory speed. The time then defines the maximum rate that data can be received. If the interrupt service routine takes longer than the time to receive the next data, there is a large risk that a data overrun will occur where data is received before the previous byte is read by the processor. To address this issue, a buffer is often used. With the later versions of the UART such as the 16450 and 16550, the FIFO buffer size has been increased. The largest buffer (16 bytes) is available on the 16550 and this device is frequently used for high speed data communications.

The 16 byte buffer means that if the processor is late for the first byte, any incoming data will simply be buffered and not cause a data overrun. As a result, the interrupt service routine need only be executed 1/16 of the times for a single buffer UART. This dramatically reduces the CPU processing needed for high speed data transfer.

There is a downside: the data now arrives in a packet with up to 16 bytes and must be processed slightly differently. With a byte at a time, the decoding of the data(i.e. is it a command or is it data that a higher level protocol may impose?) is easy to decode. With a packet of up to 16 bytes, the bytes have to be parsed to separate them out. This means that the decoding software is slightly more complex to handle both the parsing and the mechanisms to store and track the incoming data packets. An example of this in included in the chapter on buffers.

The Motorola MC68681

Within the Motorola product offering, the MC68681 has become a fairly standard UART that has been used in many MC680x0 designs. It has a quadruple buffered receiver and a double buffered transmitter. The maximum transfer rates that can be achieved are high: 9.8 Mbps with a 25 MHz clock with no clock division (×1 mode) and 612 kbps with the same clock with a divide by 16 setting (×16 mode). Each transmitter and receiver is independently programmable using one of 19 fixed rates.

It has a sophisticated interrupt structure that supports seven maskable interrupt conditions:

- Change of state on CTSx*

  This is used to support hardware handshaking. If the CTS signal changes, an interrupt can be generated to instruct the processor to stop or start sending data. This fast response coupled with the buffering ensures that data is not lost.

- Break Condition (either channel)

  The break condition is either used to request connection, i.e. send a break from a terminal to start a remote login or is symptomatic of a lost or dropped connection.

- Ready Receive/FIFO Full (either channel)

  As previously discussed, interrupts are ideal for the efficient handling and control of receive buffers. This interrupt indicates that there is data ready.

- Transmitter Ready (either channel)

  This is similar to the previous interrupt and is used to indicate that the transmitter is ready to take data for transmission.

DMA controllers

Direct memory access (DMA) controllers are frequently an elegant hardware solution to a recurring software/system problem of providing an efficient method of transferring data from a peripheral to memory.

In systems without DMA, the solution is to use the processor to either regularly poll the peripheral to see if it needs servicing or to wait for an interrupt to do so. The problem with these two methods is that they are not very efficient. Polling, by its very nature, is going to check the status and find that no action is required more times than it will find that servicing is needed. If this is not the case, then data can be lost through data over- and under-run. This means that it spends a lot of time in non-constructive work. In many embedded systems, this is not a problem but in low power systems, for example, this unnecessary work processing and power consumption cannot be tolerated.

Interrupts are a far better solution. An interrupt is sent from the peripheral to the processor to request servicing. In many cases, all that is needed is to simply empty or load a buffer. This solution starts becoming an issue as the servicing rate increases. With high speed ports, the cost of interrupting the processor can be higher than the couple of instructions that it executes to empty a buffer. In these cases, the limiting factor for the data transfer is the time to recognise, process and return from the interrupt. If the data needs to be processed on a byte by byte basis in real time, this may have to be tolerated but with high speed transfers this is often not the case as the data is treated in packets.

This is where the DMA controller comes into its own. It is a device that can initiate and control bus accesses between I/O devices and memory, and between two memory areas. With this type of facility, the DMA controller acts as a hardware implementation of the low-level buffer filling or emptying interrupt routine.

There are essentially three types of DMA controller which offer different levels of sophistication concerning memory address generation. They are often classified in terms of their addressing capability into 1D, 2D and 3D types. A 1D controller would only have a single address register, a 2D device two and a 3D device three or more.

A generic DMA controller

A generic controller consists of several components which control the operation:

* Address generator

  This is probably the most important part of a DMA controller and typically consists of a base address register and an auto-incrementing counter which increments the address after every transfer. The generated addresses are used within the actual bus transfers to access memory and/or peripherals. When a predefined number of bytes have been transferred, the base address is reloaded and the count cleared to zero ready to repeat the operation.

* Address bus

  The is where the address created by the address generator is used to access a specific memory location or peripheral.

* Data bus

  This is the data bus that is used to transfer data from the DMA controller to the destination location. In some cases, the data transfer may be made direct from the peripheral to the memory with the DMA controller directly selecting the peripheral.

* Bus requester

  This is used to request the bus from the main CPU. In older designs, the processor bus was not designed to support multiple masters and there were no bus request signals. In these cases, the processor clock was extended or delayed to steal memory cycles from the processor for the DMA controller to use.

* Local peripheral control

  This allows the DMA controller to select the peripheral and get it to accept or provide data directly or for a peripheral to request a data transfer, depending on the DMA controller's design. This is necessary to support the single or implied address mode which is explained in more detail later on.

* Interrupt signals

  Most DMA controllers can interrupt the processor when the data transfers are complete or if an error has occurred. This can prompt the processor to either reprogram the DMA controller for a different transfer or act as a signal that a new batch of data has been transferred and is ready for processing.

Operation

Using a DMA controller is reasonably simple provided the programming defines exactly the data transfer operations that the processor expects. Most errors lie in correct programming and in failing to understand how the device operates. The key phases of its operation are:

- Program the controller

  Prior to using the DMA controller, it must be configured with parameters that define the addressing such as base address and byte count that will be used to transfer the data. In addition, the device will be configured in terms of its communication with the processor and peripheral. Processor communication will normally include defining the conditions that will generate an interrupt. The peripheral communication may include defining which request pin is used by the peripheral and any arbitration mechanism that is used to reconcile simultaneous requests for DMA from two or more peripherals. The final part of this process is to define how the controller will transfer blocks of data: all at once or individually or some other combination.

Source address	FF FF 01 04
Base address	00 00 23 00
Count	00 00 00 10
Bytes transferred	00 00 00 00
Status	OK

DMA controller registers

- Start a transfer

  A DMA transfer is normally initiated in response to a peripheral request to start a transfer. It usually assumes that the controller has been correctly configured to support. With a peripheral and processor, the processor will normally request a service by asserting an interrupt pin which is connected to the processor's interrupt input(s). With a DMA controller, this peripheral interrupt signal can be used to directly initiate a transfer or if it is left attached to the processor, the interrupt service routine can start the DMA transfers by writing to the controller.

- Request the bus

  The next stage is to request the bus from the processor. With most modern processors supporting bus arbitration directly, the DMA controller issues a bus request signal to the processor which will release the bus when convenient and allow the DMA controller to proceed. Without this support, the DMA controller has to cycle steal from the processor so that it is held off the bus while the DMA controller uses it. As will be described later on in this chapter, most DMA controllers provide some flexibility concerning how they use and compete with bus bandwidth with the processor and other bus masters.

- Issue the address

  Assuming the controller has the bus, it will then issue the bus to activate the target memory location. A variety of interfaces are used — usually dependent on the number of pins that are available and include both non-multiplexed and multiplexed buses. In addition, the controller provides other signals such as read/write and strobe signals that can be used to work with the bus. DMA controllers tend to be designed for a specific processor family bus but most recent devices are also generic enough to be used with nearly any bus.

- Transfer the data

  The data is transferred either from a holding buffer within the DMA controller or directly from a peripheral.

- Update address generator

  Once the data transfer has been completed, the address generator uses the completion to calculate the address for the next transfer and update the byte/transfer counters.

- Update processor

  Depending on how the DMA controller has been programmed it can notify the processor using interrupts of events within the transfer process such as an address error or the completion of a data or block transfer.

DMA controller models

There are various modes or models that DMA controllers can support ranging from simple to complex addressing modes and single and double data transfers.

Single address model

With the single address model, the DMA controller uses its address bus to address the memory location that will participate in the bus memory cycle. The controller uses a peripheral bus — in some cases a single select and a read/write pin — to select the peripheral device so its data bus becomes active. The select signal from the processor often has to generate an address to access the specific register within the peripheral such as the buffer register. If the peripheral is prompting the transfer, the peripheral would pull down a request line — typically its interrupt line is used for this purpose.

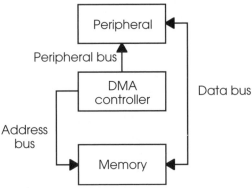

Single address or implicit address mode

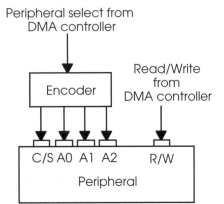

Activating the peripheral by the DMA controller

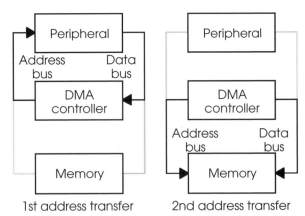

1st address transfer   2nd address transfer

Dual address transfer

In this way, data can be transferred between the memory and peripheral as needed, without the data being transferred through the DMA controller and thus taking two cycles. This model is also known as the implicit address because the second address is implied and not directly given, i.e. there is no source address supplied.

Dual address model

The dual address mode uses two addresses and two accesses to transfer data between a peripheral or memory and another memory location. This consumes two bus cycles and uses a buffer within the DMA controller to temporarily hold data.

1D model

The 1D model uses an address location and a counter to define the sequence of addresses that are used during the DMA cycles. This effectively defines a block of memory which is used for the access. The disadvantage of this arrangement is that when the block is transferred, the address and counter are usually reset automatically and thus can potentially overwrite the previous data. This can be prevented by using an interrupt from the DMA controller to the processor when the counter has expired. This allows the CPU the opportunity to change the address so that next memory block to be used is different.

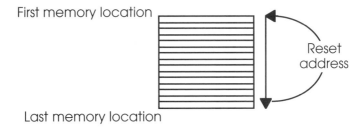

First memory location   Reset address

Last memory location

A circular buffer

This model left on its own can be used to implement a circular buffer where the automatic reset is used to bring the address back to the beginning. Circular buffering can be an efficient technique in terms of both the size of buffering and timing constraints.

2D model

While the 1D model is simple, there are times especially with high speed data where the addressing mode is not powerful enough even though it can be augmented through processor intervention. A good example of this is with packet-based communication protocols where the data is wrapped up with additional information in the form of headers. The packets typically have a maximum or fixed data format and thus large amounts of consecutive data has to be split and header and trailer information either added or removed.

With the 2D model, an address stride can be specified which is used to calculate an offset to the base address at the end of a count. This allows DMA to occur in non-consecutive blocks of memory. Instead of the base address being reset to the original address, it has the stride added to it. In addition the count register is normally split into two: one register to specify the count for the block and a second register to specify the total number of blocks or bytes to be transferred. Using these new features, it is easy to set up a DMA controller to transfer data and split into blocks ready for the insertion of header information. The diagram shows how this can be done.

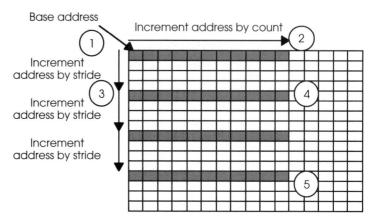

1. Use base address to start transfer. Increment until counter expires.

2. Reset counter and change base address using stride.

3. Use base address and increment until counter expires.

4. Reset counter and change base address using stride.

5. Repeat until total number of requested bytes transferred. (Total count = 11 x 4 = 44 bytes)

2D addressing structure

3D model

The third type of controller takes the idea of address strides a step further by defining the ability to change the stride automatically so that blocks of different sizes and strides can be created. It is possible to simulate this with a 2D controller and software so that the processor reprograms the device to simulate the automatic change of stride.

Channels and control blocks

By now, it should be reasonably clear that DMA controllers need to be pre-programmed with a block of parameters to allow them to operate. The hardware interface that they use is common to almost every different set of parameters — the only real difference is when a single or dual address mode is used with the need to directly access a peripheral as well as the memory address bus.

It is also common for a peripheral to continually use a single set of parameters. As a result, the processor has to continually re-program the DMA controller prior to use if it is being shared between several peripherals. Each peripheral would have to interrupt the processor prior to use of the DMA to ensure that it was programmed. Instead of removing the interrupt burden from the processor, the processor still has it — albeit it is now programming the DMA controller and not moving data. Moving data could even be a lighter load!

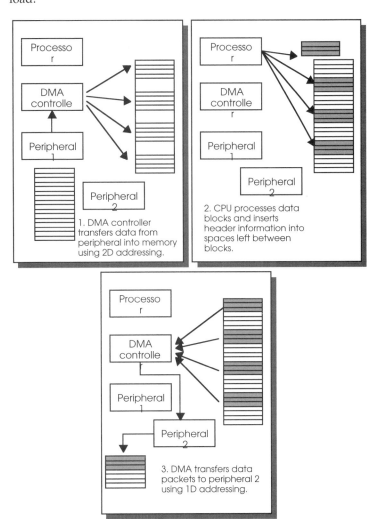

Using 2D addressing to create space for headers

To overcome this the idea of channels of control blocks was developed. Here the registers that contain the parameters are duplicated with a set for each channel. Each peripheral is assigned an external request line which when asserted will cause the DMA controller to start a DMA transfer in accordance with the parameters that have been assigned with the request line. In this way, a single DMA controller can be shared with multiple peripherals with each peripheral having its own channel. This is how the DMA controller in the IBM PC works. It supports four channels (0 to 3).

DMA control block #1

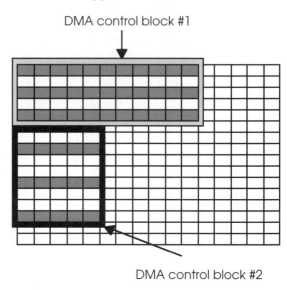

DMA control block #2

Using control blocks

An extension to this idea of channels, or control blocks as they are also known, is the idea of chaining. With chaining, channels are linked together so that more complex patterns can be created. The first channel controls the DMA transfers until it has completed its allotted transfers and then control is passed to the next control block that has been chained to it. This simple technique allows very complex addressing patterns to be created such as described in the paragraphs on 3D models.

There is one problem with the idea of channels and external pins: what happens if multiple requests are received by the DMA controller at the same time? To resolve this situation, arbitration is used to prioritise multiple requests. This may be a strict priority scheme where one channel has the highest priority or can be a fairer system such as a round-robin where the priority is equally distributed to give a fairer allocation of priority.

Sharing bus bandwidth

The DMA controller has to compete with the processor for external bus bandwidth to transfer data and as such can affect the processor's performance directly. With processors that do not have any cache or internal memory, such as the 80286 and the MC68000, their bus utilisation is about 80–95% of the bandwidth and therefore any delay in accessing external memory will result in a decreased processor performance budget and potentially longer interrupt latency — more about this in the chapter on interrupts.

For devices with caches and/or internal memory, their external bus bandwidth requirements are a lot lower and thus the DMA controller can use bus cycles without impeding the processor's performance. This last statement depends on the chances of the DMA controller using the bus at the same time as the processor. This in turn depends on the frequency and size of the DMA transfers. To provide some form of flexibility for the designer so that a suitable trade-off can be made, most DMA controllers support different types of bus utilisation.

- Single transfer

  Here the bus is returned back to the processor after every transfer so that the longest delay it will suffer in getting access to memory will be a bus cycle.

- Block transfer

  Here the bus is returned back to the processor after the complete block has been sent so that the longest delay the processor will suffer in will be the time of a bus cycle multiplied by the number of transfers to move the block. In effect, the DMA controller has priority over the CPU in using the bus.

- Demand transfer

  In this case, the DMA controller will hold the bus for as long as an external device requests it to do so. While the bus is held, the DMA controller is at liberty to transfer data as and when needed. In this respect, there may be gaps when the bus is retained but no data is transferred.

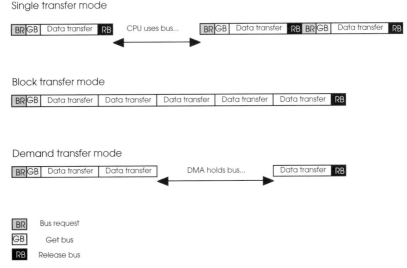

DMA transfer modes

DMA implementations

Intel 8237

This device is used in the IBM PC and is therefore probably the most used DMA controller in current use. Like most peripherals today, it has moved from being a separate entity to be part of the PC chip set that has replaced the 100 or so devices in the original design with a single chip.

It can support four main transfer modes including single and block transfers, the demand mode and a special cascade mode where additional 8237 DMA controllers can be cascaded to expand the four channels that a single device can support. It can transfer data between a peripheral and memory and by combining two channels together, perform memory to memory transfers although this is not used or supported within the IBM PC environment. In addition, there is a special verify transfer mode which is used within the PC to generate dummy addresses to refresh the DRAM memory. This is done in conjunction with a 15 μs interrupt derived from a timer channel on the PC motherboard.

To resolve simultaneous DMA requests, there is an internal arbitration scheme which supports either a fixed or rotating priority scheme.

Motorola MC68300 series

Whereas five or 10 years ago, DMA controllers were freely available as separate devices, the increasing ability to integrate functionality has led to their demise as separate entities and most DMA controllers are either integrated onto the peripheral or as in this case onto the processor chip. The MC68300 series combine an MC68000/ MC68020 type of processor with peripherals and DMA controllers.

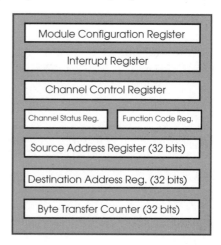

MC683xx generic DMA controller

It consists of a two channel fully programmable DMA controller that can support high speed data transfer rates of 12.5 Mbytes/s in dual address transfer mode or 50.0 Mbytes/s in single address mode at a 25 MHz clock rate. The dual address mode is considerably slower because two cycles have to be performed as previously described. By virtue of its integration onto the processor chip with the peripherals and internal memory, it can DMA data between internal and external resources. Internal cycles can be programmed to occupy 25, 50, 75, or 100% of the available internal bus bandwidth while external cycles support burst and single transfer mode.

The source and destination registers can be independently programmed to remain constant or incremented as required.

Using another CPU with firmware

This is a technique that is sometimes used where a DMA controller is not available or is simply not fast or sophisticated enough. The DMA CPU requires its own local memory and program so that it can run in isolation and not burden the main memory bus. The DMA CPU is sent messages which instruct it on how to perform its DMA operations. The one advantage that this offers is that the CPU can be programmed with higher level software which can be used to process the data as well as transfer it. Many of the processors used in embedded systems fall into this category of device.

5   Interfacing to the analogue world

This chapter discusses the techniques used to interface to the outside world which unfortunately is largely analogue in nature. It discusses the process of analogue to digital conversion and basic power control techniques to drive motors and other similar devices from a microcontroller.

Analogue to digital conversion techniques

The basic principle behind analogue to digital conversion is simple and straightforward: the analogue signal is sampled at a regular interval and each sample is divided or quantised by a given value to determine the number of given units of value that approximate to the analogue value. This number is the digital equivalent of the analogue signal.

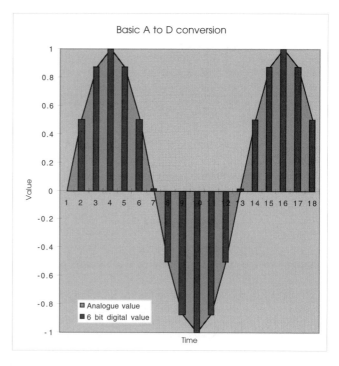

Basic A to D conversion

The combination graph shows the general principle. The grey curve represents an analogue signal which, in this case, is a sine wave. For each cycle of the sine wave, 13 digital samples are taken which encode the digital representation of the signal.

Quantisation errors

Careful examination of the combination chart reveals that all is not well. Note the samples at time points 7 and 13. These should be zero — however, the conversion process does not convert them to zero but to a slightly higher value. The other points show similar errors; this is the first type of error that the conversion process can cause. These errors are known as quantisation errors and are caused by the fact that the digital representation is step based and consists of a selection from 1 of a fixed number of values. The analogue signal has an infinite range of values and the difference between the digital value the conversion process has selected and the analogue value is the quantisation error.

Size	Resolution	Storage (1s)	Storage (60s)	Storage (300s)
4 bit	0.0625000000	22050	1323000	6615000
6 bit	0.0156250000	33075	1984500	9922500
8 bit	0.0039062500	44100	2646000	13230000
10 bit	0.0009765625	55125	3307500	16537500
12 bit	0.0002441406	66150	3969000	19845000
16 bit	0.0000152588	88200	5292000	26460000
32 bit	0.0000000002	176400	10584000	52920000

Resolution assumes an analogue value range of 0 to 1
Storage requirements are in bytes and a 44.1 kHz sample rate

Digital bit size, resolution and storage

The size of the quantisation error is dependent on the number of bits used to represent the analogue value. The table shows the resolution that can be achieved for various digital sizes. As the table depicts, the larger the digital representation, the finer the analogue resolution and therefore the smaller the quantisation error and resultant distortion. However, the table also shows the increase which occurs with the amount of storage needed. It assumes a sample rate of 44.1 kHz, which is the same rate as used with an audio CD. To store 5 minutes of 16 bit audio would take about 26 Mbytes of storage. For a stereo signal, it would be twice this value.

Sample rates and size

So far, much of the discussion has been on the sample size. Another important parameter is the sampling rate. The sampling rate is the number of samples that are taken in a time period, usually one second, and is normally measured in hertz, in the same way that frequencies are measured. This determines several aspects of the conversion process:

• It determines the speed of the conversion device itself. All converters require a certain amount of time to perform the conversion and this conversion time determines the maximum rate at which samples can be taken. Needless to say, the fast converters tend to be the more expensive ones.

- The sample rate determines the maximum frequency that can be converted. This is explained later in the section on Nyquist's theorem.
- Sampling must be performed on a regular basis with exactly the same time period between samples. This is important to remove conversion errors due to irregular sampling.

Irregular sampling errors

The line chart shows the effect of irregular sampling. It effectively alters the amplitude or magnitude of the analogue signal being measured. With reference to the curve in the chart, the following errors can occur:

- If the sample is taken early, the value converted will be less than it should be. Quantisation errors will then be added to compound the error.
- If the sample is taken late, the value will be higher than expected. If all or the majority of the samples are taken early, the curve is reproduced with a similar general shape but with a lower amplitude. One important fact is that the sampled curve will not reflect the peak amplitudes that the original curve will have.
- If there is a random timing error — often called jitter — then the resulting curve is badly distorted, again as shown in the chart.

Other sample rate errors can be introduced if there is a delay in getting the samples. If the delay is constant, the correct characteristics for the curve are obtained but out of phase. It is interesting that there will always be a phase error due to the conversion time taken by the converter. The conversion time will delay the digital output and therefore introduces the phase error — but this is usually very small and can typically be ignored.

The phase error shown assumes that all delays are consistent. If this is not the case, different curves can be obtained as shown in the next chart. Here the samples have been taken at random and at slightly delayed intervals. Both return a similar curve to that of the original value — but still with significant errors.

In summary, it is important that samples are taken on a regular basis with consistent intervals between them. Failure to observe these design conditions will introduce errors. For this reason, many microprocessor-based implementations use a timer and interrupt service routine mechanism to gather samples. The timer is set-up to generate an interrupt to the processor at the sampling rate frequency. Every time the interrupt occurs, the interrupt service routine reads the last value for the converter and instructs it to start a new conversion before returning to normal execution. The instructions always take the same amount of time and therefore sampling integrity is maintained.

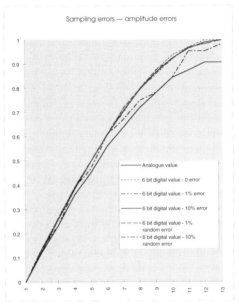

Sampling errors — amplitude errors

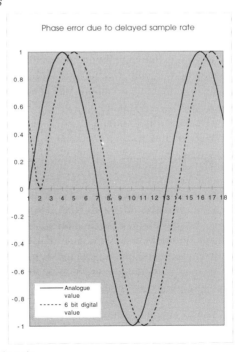

Phase errors due to delayed sample rate

Nyquist's theorem

The sample rate also must be chosen carefully when considering the maximum frequency of the analogue signal being converted. Nyquist's theorem states that the *minimum* sampling rate frequency should be twice the maximum frequency of the analogue signal. A 4 kHz analogue signal would need to be sampled at twice that frequency to convert it digitally. For example, a hi-fi audio signal with a frequency range of 20 to 20 kHz would need a minimum sampling rate of 40 kHz.

Higher frequency sampling introduces a frequency component which is normally filtered out using an analogue filter.

Codecs

So far the discussion has been based on analogue to digital (A to D) and digital to analogue (D to A) converters. These are the names used for generic converters. Where both A to D and D to A conversion is supported, they can also be called codecs. This name is derived from **co**der-**dec**oder and is usually coupled with the algorithm that is used to perform the coding. Generic A to D conversion is only one form of coding; many others are used within the industry where the analogue signal is converted to the digital domain and then encoded using a different technique. Such codecs are often prefixed by the algorithm used for the encoding.

Linear

A linear codec is one that is the same as the standard A to D and D to A converters so far described, i.e. the relationship between the analogue input signal and the digital representation is linear. The quantisation step is the same throughout the range and thus the increase in the analogue value necessary to increment the digital value by one is the same, irrespective of the analogue or digital values. Linear codecs are frequently used for digital audio.

A-law and μ-law

For telecommunications applications with a limited bandwidth of 300 to 3100 Hz, logarithmic codecs are used to help improve quality. These codecs, which provide an 8 bit sample at 8 kHz, are used in telephones and related equipment. Two types are in common use: the a-law codec in the UK and the μ-law codec in the US. By using a logarithmic curve for the quantisation, where the analogue increase to increment the digital value varies depending on the size of the analogue signal, more digital bits can be allocated to the more important parts of the analogue signal and thus improve their resolution. The less important areas are given less bits and, despite having coarser resolution, the quality reduction is not really noticeable because of the small part they contribute to the signal. Conversion between a linear digital signal and a-law/μ-law or between an a-law and μ-law signal is easily performed using a look-up table.

PCM

The linear codecs that have been so far described are also known as PCM — pulse code modulation codecs. This comes from the technique used to reconstitute the analogue signal by supplying a series of pulses whose amplitude is determined by the digital value. This term is frequently used within the telecommunications industry.

There are alternative ways of encoding using PCM which can reduce the amount of data needed or improve the resolution and accuracy.

DPCM

Differential pulse coded modulation (DPCM) is similar to PCM, except that the value encoded is the difference between the current sample and the previous sample. This can improve the accuracy and resolution by having a 16 bit digital dynamic range without having to encode 16 bit samples. It works by increasing the dynamic range and defining the differential dynamic range as a partial value of it. By encoding the difference, the smaller digital value is not exceeded but the overall value can be far greater. There is one proviso: the change in the analogue value from one sample to another must be less than the differential range and this determines the maximum slope of any waveform that is encoded. If the range is exceeded, errors are introduced.

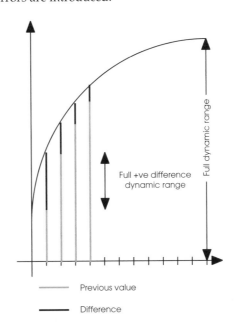

DPCM encoding

The diagram shows how this encoding works. The analogue value is sampled and the previous value subtracted. The result is then encoded using the required sample size and allowing for a plus and minus value. With an 8 bit sample size, 1 bit is used as a sign bit and the remaining 7 bits are used to encode data. This allows the previous value to be used as a reference, even if the next value is smaller. The 8 bits are then stored or incorporated into a bitstream as with a PCM conversion.

To decode the data, the reverse operation is performed. The signed sample is added to the previous value, giving the correct digital value for decoding. In both the decode and encode process, values which are far larger than the 8 bit sample are stored. This type of encoding is easily performed with a microprocessor with 8 bit data and a 16 bit or larger accumulator.

A to D and D to A converters do not have to cope with the full resolution and can simply be 8 bit decoders. These can be used provided analogue subtractors and adders are used in conjunction with them. The subtractor is used to reduce the analogue input value before inputting to the small A to D converter. The adder is used to create the final analogue output from the previous analogue value and the output from the D to A converter.

ADPCM

Adaptive differential pulse code modulation (ADPCM) is a variation on the previous technique and frequently used in telecommunications. The difference is encoded as before but instead of using all the bits to encode the difference, some bits are used to encode the quantisation value that was used to encode the data. This means that the resolution of the difference can be adjusted — adapted — as needed and, by using non-linear quantisation values, better resolution can be achieved and a larger dynamic range supported.

Power control

Most embedded designs need to be able to switch power in some way or another, if only to drive an LED or similar indicator. This, on first appearances, appears to be quite simple to do but there are some traps that can catch designers out. This section goes through the basic principles and techniques.

Matching the drive

The first problem that faces any design is matching the logic level voltages with that of a power transistor or similar device. It is often forgotten or assumed that with logic devices, a logical high is always 5 volts and that a logical low is zero. A logical high of 5 volts is more than enough to saturate a bipolar transistor and turn it on. Similarly, 0 volts is enough to turn off such a transistor.

Unfortunately, the specifications for TTL compatible logic levels are not the same as indicated by these assumptions. The voltage levels are define a logic low output as any voltage below a maximum which is usually 0.4 volts and a logic high output as a voltage above 2.4 volts assuming certain bus capacitance and load currents and a supply voltage of 4.5 to 5.5 volts. These figures are typical and can vary.

If the output high is used to drive a bipolar transistor, then the 2.4 volt value is high enough to turn on the transistor. The only concern is the current drive that the output can provide. This value times the gain of the transistor determines the current load that the transistor can provide. With an output low voltage of 0.4 volts, the situation is less clear and is dependent on the biasing used on the transistor. It is possible that instead of turning the transistor off completely, it partially turns the device off and some current is still provided.

With CMOS logic levels, similar problems can occur. Here the logic high is typically two thirds of the supply voltage or higher and a logic low is one third of the supply voltage or lower. With a 5 volt supply, this works out at 3.35 volts and 1.65 volts for the high and low states. In this case, the low voltage is above the 0.7 volts needed to turn on a transistor and thus the transistor is likely to be switched on all the time irrespective of the logic state. These voltage mismatches can also cause problems when combining CMOS and TTL devices using a single supply. With bipolar transistors there are several techniques that can be used to help avoid these problems:

- Use a high gain transistor

  The higher the gain of the transistor, the lower the drive needed from the output pin and the harder the logic level will be. If the required current is high, then the voltage on the output is more likely to reach its limits. With an output high, it will fall to the minimum value. With an output low, it will rise to the maximum value.

  Darlington transistor pairs are often used because they have a far higher gain compared to a single transistor.

- Use a buffer pack

  Buffer packs are logic devices that have a high drive capability and can provide higher drive currents than normal logic outputs. This increased drive capability can be used to drive an indicator directly or can be further amplified.

- Use a field effect transistor (FET)

  These transistors are voltage controlled and have a very high effective gain and thus can be used to switch heavy loads easily from a logic device. There are some problems, however, in that the gate voltages are often proportions of the supply voltages and these do not match with the logic voltage levels that are available. As a result, the FET does not switch correctly. This problem has been solved by the introduction of logic level switching FETs that will switch using standard logic voltages. The advantage that these offer is that they can simply have their gate directly connected to the logic output. The power supply and load are connected through the FET which acts as a switch.

Using H bridges

Using logic level FETs is a very simple and effective way of providing DC power control. With the FET acting like a power switch whose state reflects the logic level output from the digital controller, it is possible to combine several switches to create H bridges which allow a DC motor to be switched on and reversed in direction. This is done by using two outputs and four FETs acting as switches.

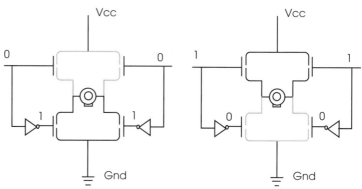

Switching the motor off using an H bridge

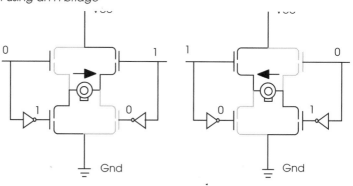

Switching the motor on with an H bridge

The FETs are arranged in two pairs so that by switching one on and the other off, one end of the motor can be connected to ground (0 volts) or to the voltage supply Vcc. Each FET in the pair is driven from a common input signal which is inverted on its way to one of the FETs. This ensures that only one of the pairs switches on in response to the input signal. With the two pairs, two input signals are needed. When these signals are the same, i.e. 00 or 11, either the top or bottom pairs of FETS are switched on and no voltage differential is applied across the motor, so nothing happens. This is shown in the first diagram where the switched-on paths are shown in black and the switched-off paths are in grey.

If the input signals are different then a top and a bottom FET is switched on and the voltage is applied across the motor and it revolves. With a 01 signal it moves in one direction and with a 10 signal it moves in the reverse direction.

This type of bridge arrangement is frequently used for controlling DC motors or any load where the voltage may need reversing.

Driving LEDs

Light emitting diodes (LEDs) are often used as indicators in digital systems and in many cases can simply be directly driven from a logic output provided there is sufficient current and voltage drive.

The voltage drive is necessary to get the LED to illuminate in the first place. LEDs will only light up when their diode reverse breakdown voltage is exceeded. This is usually about 2 to 2.2 volts and less than the logic high voltage. The current drive determines how bright the LED will appear and it is usual to have a current limiting resistor in series with the LED to prevent it from drawing too much current and overheating. For a logic device with a 5 volt supply a 300 Ω resistor will limit the current to about 10 mA. The problem comes if the logic output is only 2.4 or 2.5 volts and not the expected 5 volts. This means that the resistor is sufficient to drop enough voltage so that the LED does not light up. The solution is to use a buffer so that there is sufficient current drive or alternatively use a transistor to switch on the LED. There are special LED driver circuits packs available that are designed to connect directly to an LED without the need for the current limiting resistor. The resistor or current limiting circuit is included inside the device.

Interfacing to relays

Another method of switching power is to use a mechanical relay where the logic signal is used to energise the relay. The relay contacts make or break accordingly and switch the current. The advantage of a relay is that it can be used to switch either AC or DC power and there is no electrical connections between the low power relay coil connected to the digital circuits and the power load that is being switched. As a result, they are frequently used to switch high loads.

Relays do suffer from a couple of problems. The first is that the relay can generate a back voltage across its terminals when the energising current is switched off, i.e. when the logic output switches from a high to a low. This back EMF as it is known can be a high voltage and cause damage to the logic circuits. A logic output does not expect to see an input voltage differential of several tens of volts! The solution is to put a diode across the relay circuits so that in normal operation, the diode is reverse biased and does nothing. When the back EMF is generated, the diode starts to conduct and the voltage is shorted out and does no damage. This problem is experienced with any coil, including those in DC motors. It is advisable to fit a diode when driving these components as well.

The other problem is that the switch contacts can get sticky where they are damaged with the repeated current switching. This can erode the contacts and cause bad contacts or in some cases can cause local overheating so that the contacts weld themselves together. The relay is now sticky in that the contacts will not change when the coil is de-energised.

Interfacing to DC motors

So far with controlling DC motors, the emphasis has been simple on-off type switching. It is possible with a digital system to actually provide speed control using a technique called pulse width modulation.

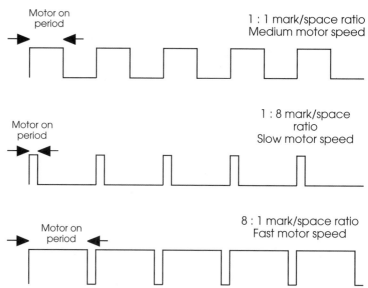

Using different PWM waveforms to control a
DC motor speed

With a DC motor, there are two techniques for controlling the motor speed: the first is to reduce the DC voltage to the motor. The higher the voltage, the faster it will turn. At low voltages, the control can be a bit hit and miss and the power control is inefficient. The alternative technique called pulse width modulation (PWM) will control a motor speed not by reducing the voltage to the motor but by reducing the time that the motor is switched on.

This is done by generating a square wave at a frequency of several hundred hertz and changing the mark/space ratio of the wave form. With a large mark and a low space, the voltage is applied to the motor for almost all of the cycle time, and thus the motor will rotate very quickly. With a small mark and a large space, the opposite is true. The diagram shows the waveforms for medium, slow and fast motor control.

The only difference between this method of control and that for a simple on-off switch is the timing of the pulses from the digital output to switch the motor on and off. There are several methods that can be used to generate these waveforms.

Software only

With a software-only system, the waveform timing is done by creating some loops that provide the timing functions. The program pseudo code shows a simple structure for this. The first action is to switch the motor on and then to start counting through a delay loop. The length of time to count through the delay loop determines the motor-on period. When the count is finished, the motor is switched off. The next stage is to count through a second delay loop to determine the motor-off period.

```
repeat (forever)
{
    switch on motor
    delay loop1
    switch off motor
    delay loop2
}
```

This whole procedure is repeated for as long as the motor needs to be driven. By changing the value of the two delays, the mark-space ratio of the waveform can be altered. The total time taken to execute the repeat loop gives the frequency of the waveform.

This method is processor intensive in that the program has to run while the motor is running. On first evaluation, it may seem that while the motor is running, nothing else can be done. This is not the case. Instead of simply using delay loops, other work can be inserted in here whose duration now becomes part of the timing for the PWM waveform. If the work is short, then the fine control over the mark/space ratio is not lost because the contribution that the work delay makes compared to the delay loop is small. If the work is long, then the minimum motor-on time and thus motor speed is determined by this period.

```
repeat (forever)
{
    switch on motor
```

```
        perform task a
        delay loop1
        switch off motor
        delay loop2
}
```

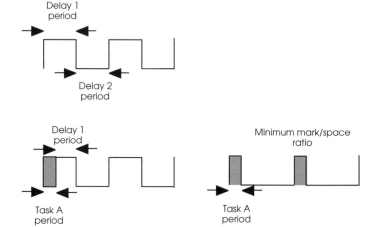

The timing diagrams for the software PWM implementation

The timing diagrams for the software loop PWM waveforms are shown in the diagrams above. In general, software only timing loops are not efficient methods of generating PWM waveforms for motor control. The addition of a single timer greatly improves the mechanism.

Using a single timer

By using a single timer, PWM waveforms can be created far easier and free up the processor to do other things without impacting the timing. There are several methods that can be used to do this. The key principle is that the timer can be programmed to create a periodic interrupt.

Method 1 — using the timer to define the on period

With this method, the timer is used to generate the on period. The processor switches the motor on and then starts the timer to count down. While the timer is doing this, the processor is free to do what ever work is needed. The timer will eventually time out and generate a processor interrupt. The processor services the interrupt and switches the motor off. It then goes into a delay loop still within the service routine until the time period arrives to switch the motor on again. The processor switches the motor on, resets the timer and starts it counting and continues with its work by returning from the interrupt service routine.

Method — using the timer to define frequency period

With this method, the timer is used to generate a periodic interrupt whose frequency is set by the timer period. When the processor services the interrupt, it uses a software loop to determine the on period. The processor switches on the motor and uses the software delay to calculate the on period. When the delay loop is completed, it switches off the motor and can continue with other work until the timer generates the next interrupt.

Method 3 — using the timer to define both the on and off periods

With this method, the timer is used to generate both the on and off periods.

The processor switches the motor on, loads the timer with the on-period value and then starts the timer to count down. While the timer is doing this, the processor is free to do what ever work is needed. The timer will eventually time out and generate a processor interrupt, as before. The processor services the interrupt and switches the motor off. It then loads the timer with the value for the off period. The processor then starts the timer counting and continues with its work by returning from the interrupt service routine.

The timer now times out and generates an interrupt. The processor services this by switching the motor on, loading the timer with the one delay value and setting the timer counting before returning from the interrupt.

As a result, the processor is only involved when interrupted by the timer to switch the motor on or off and load the timer with the appropriate delay value and start it counting. Of all these three methods, this last method is the most processor efficient. With methods 1 and 2, the processor is only free to do other work when the mark/space ratio is such that there is time to do it. With a long motor-off period, the processor performs the timing in software and there is little time to do anything else. With a short motor-off period, there is more processing time and far more work can be done. The problem is that the work load that can be achieved is dependent on the mark/space ratio of the PWM waveform and engine speed. This can be a major restriction and this is why the third method is most commonly used.

Using multiple timers

With two timers, it is possible to generate PWM waveforms with virtually no software intervention. One timer is set-up to generate a periodic output at the frequency of the required PWM waveform. This output is used to trigger a second timer which is configured as a monostable. The second timer output is used to provide the motor-on period. If these timers are set to automatically reload, the first timer will continually trigger the second and thus generate a PWM waveform. By changing the delay value in the second timer, the PWM mark/space ratio can be altered as needed.

6 Interrupts and exceptions

Interrupts are probably the most important aspect of any embedded system design and potentially can be responsible for many problems when debugging a system. Although they are simple in concept, there are many pitfalls that the unwary can fall into. This chapter goes through the principles behind interrupts, the different mechanisms that are used with various processor architectures and provides a set of do's and don'ts to help guide the designer.

What is an interrupt?

An interrupt is an event from either an internal or external source where a processor will stop its current processing and switch to a different instruction sequence in response to an event that has occurred either internally or externally. The advantage of the interrupt is that it allows the designer to split software into two types: background work where tasks are performed while waiting for an interrupt and foreground work where tasks are performed in response to interrupts. The interrupt mechanism is normally transparent to the background software and it is not aware of the existence of the foreground software. As a result, it allows software and systems to be developed in a modular fashion without having to create a spaghetti bolognese blob of software where all the functions are thrown together. The best way of explaining this is to consider several alternative methods of writing software for a simple system.

The system consists of a processor that has to periodically read in data from a port, process it and write it out. While waiting for the data, it is designed to perform some form of statistical analysis.

The spaghetti method

In this case, the code is written in a straight sequence where occasionally the analysis software goes and polls the port to see if there is data present. If there is data present, this is processed before returning to the analysis. To write such code, there is extensive use of branching to effectively change the flow of execution from the background analysis work to the foreground data transfer operations. The periodicity is controlled by two factors:

- The number of times the port is polled while executing the analysis task. This is determined by the data transfer rate.
- The time taken between each polling operation to execute the section of the background analysis software.

With a simple system, this is not too difficult to control but as the complexity increases or the data rates go up requiring a higher polling rate, this software structure rapidly starts to fall about and become inefficient. The timing is software based and therefore will change if any of the analysis code is changed or extended.

If additional analysis is done, then more polling checks need to be inserted. As a result, the code often quickly becomes a hard to understand mess.

The situation can be improved through the use of subroutines so that instead of reproducing the code to poll and service the ports, subroutines are called and while this does improve the structure and quality of the code, it does not remove the fundamental problem of a software timed design.

There are several difficulties with this type of approach:

- The system timing and synchronisation is completely software dependent which means that it now assumes certain processor speeds and instruction timing to provide a required level of performance.

- If the external data transfers are in bursts and they are asynchronous, then the polling operations are usually inefficient. A large number of checks will be needed to ensure that data is not lost. This is the old polling vs. interrupt argument reappearing.

- It can be very difficult to debug because there are multiple element/entry points within the code that perform the same operation. As a result, there are two asynchronous operations going on in the system. The software execution and asynchronous incoming data will mean that the routes from the analysis software to the polling and data transfer code will be used almost at random. The polling/data transfer software that is used will depend on when the data arrived and what the background software was doing. In this way, it makes reproducing errors extremely difficult to reproduce and frequently can be responsible for intermittent errors.

- The software/system design is now time referenced as opposed to being event driven. For the system to work, there are time constraints imposed on it such as the frequency of polling which cannot be broken. As a result, the system can become very inefficient. To use an office analogy, it is not very efficient to have to send a nine page fax if you have to be present to insert each page separately. You either stay and do nothing while you wait for the right moment to insert the next page or you have to check the progress repeatedly so that you do not miss the next slot.

Using interrupts

An interrupt is, as its name suggests, a way of stopping the current software thread that the processor is executing, changing to a different software routine and executing it before restoring the processor's status to that prior to the interrupt so that it can continue processing.

Interrupts can happen asynchronously to the operation and therefore can be used very efficiently with systems that are event as opposed to time driven. However, they can used to create time driven systems without having to resort to software-based timers.

To convert the previous example to one using interrupts, all the polling and data port code is removed from the background analysis software. The data transfer code is written as part of the interrupt service routine (ISR) associated with the interrupt generated by the data port hardware. When the port receives a byte of data, it generates an interrupt. This activates the ISR which processes the data before handing execution back to the background task. The beauty of this type of operation is that the background task can be written independently of the data port code and that the whole timing of the system is now moved from being dependent on the polling intervals to one of how quickly the data can be accessed and processed.

Interrupt sources

There are many sources for interrupts varying from simply asserting an external pin to error conditions within the processor that require immediate attention.

Internal interrupts

Internal interrupts are those that are generated by on-chip peripherals such as serial and parallel ports. With an external peripheral, the device will normally assert an external pin which is connected to an interrupt pin on the processor. With internal peripherals, this connection is already made. Some integrated processors allow some flexibility concerning these hardwired connections and allow the priority level to be adjusted or even masked out or disabled altogether.

External interrupts

External interrupts are the common method of connecting external peripherals to the processor. They are usually provided through external pins that are connected to peripherals and are asserted by the peripheral. For example, a serial port may have a pin that is asserted when there is data present within its buffers. The pin could be connected to the processor interrupt pin so that when the processor sees the data ready signal as an interrupt. The corresponding interrupt service routine would then fetch the data from the peripheral before restoring the previous processing.

Exceptions

Many processor architectures use the term exception as a more generic term for an interrupt. While the basic definition is the same — an event that changes the software flow and to process the event — an exception is extended to cover any event including internal and external interrupts that causes the processor to change to a service routine. Typically, exception processing is normally coupled with a change in the processor's mode. This will be described in more detail for some example processors later in this chapter.

The range of exceptions can be large and varied. A MC68000 has a 256 entry vector table which describes about 90 exception conditions with the rest reserved for future expansion. An 8 bit micro may have only a few.

Software interrupts

The advantage of an interrupt is that it includes a mechanism to change the program flow and in some processor architectures, to change into a more protected state. This means that an interrupt could be used to provide an interface to other software such as an operating system. This is the function that is provided by the software interrupt. It is typically an instruction or set of instructions that allows a currently executing software sequence to change flow and return using the more normal interrupt mechanism. With devices like the Z80 this function is provided by the SWI (software interrupt instruction). With the MC68000 and PowerPC architectures, the TRAP instruction is used.

To use software interrupts efficiently, additional data to specify the type of request and/or data parameters has to be passed to the specific ISR that will service the software interrupt. This is normally done by using one or more of the processor's registers. The registers are accessible by the ISR and can be used to pass status information back to the calling software.

It could be argued that there is no need to use software interrupts because branching to different software routines can be achieved by branches and jumps. The advantage that a software interrupt offers is in providing a bridge and routine between software running in the normal user mode and other software running in a supervisor mode. The different modes allow the resources such as memory and associated code and data to be protected from each other. This means that if the user causes a problem or makes an incorrect call, then the supervisor code and data are not at risk and can therefore survive and thus have a chance to restore the system or at least shut it down in an orderly manner.

Non-maskable interrupts

A non-maskable interrupt (NMI) is as its name suggests an external interrupt that cannot be masked out. It is by default at the highest priority of any interrupt and will always be recognised and processed. In terms of a strict definition, it is masked out when the ISR starts to process the interrupt so that it is not repeatedly recognised as a separate interrupt and therefore the non-maskable part refers to the ability to mask the interrupt prior to its assertion.

The NMI is normally used as a last resort to generate an interrupt to try and recover control. This can be presented as either a reset button or connected to a fault detection circuit such as a memory parity or watchdog timer. The 80x86 NMI as used on the IBM PC is probably the most known implementation of this function. If the PC memory subsystem detects a parity error, the parity circuitry asserts the NMI. The associated ISR does very little except stop the processing and flash up a window on the PC saying that a parity error has occurred and please restart the machine.

Recognising an interrupt

The start of the whole process is the recognition of an interrupt. Internal interrupts are normally defined by the manufacturer and are already hardwired. External interrupts, however, are not and can use a variety of mechanisms.

Edge triggered

With the edge triggered interrupt, it is the clock edge that is used to generate the interrupt. The transition can either be from a logical high to low or vice versa. With these systems, the recognition process is usually in two stages. The first stage is the external transition that typically latches an interrupt signal. This signal is then checked on an instruction boundary and, if set, starts the interrupt process. At this point, the interrupt has been successfully recognised and the source removed.

Level triggered

With a level triggered interrupt, the trigger is dependent on the logic level. Typically, the interrupt pin is sampled on a regular basis, e.g. after every instruction or on every clock edge. If it is set to the appropriate logic level, the interrupt is recognised and acted upon. Some processors require the level to be held for a minimum number of clocks or for a certain pulse width so that extraneous pulses that are shorter in duration than the minimum pulse width are ignored.

Maintaining the interrupt

So far, the recognition of an interrupt has concentrated on simply asserting the interrupt pin. This implies that provided the minimum conditions have been met, the interrupt source can be removed. Many microprocessor manufacturers recommend that this is not done and that the interrupt should be maintained until it has been explicitly serviced and the source instructed to remove it.

Internal queuing

This last point also raises a further potential complication. If an interrupt is asserted so that it conforms with the recognition conditions, removed and reasserted, the expectation would be that the interrupt service routine would be executed twice to service each interrupt. This assumes that there is an internal counter within the processor that can count the number of interrupts and thus effectively queue them. While this might be expected, this is not the case with most processors. The first interrupt would be recognised and, until it is serviced, all other interrupts generated using the pin are ignored. This is one reason why many processors insist on the maintain until serviced approach with interrupts. Any subsequent interrupts that have the same level will be maintained after the first one has been serviced and its signal removed. When the exception processing is completed, the remaining interrupts will be recognised and processed one by one until they are all serviced.

The interrupt mechanism

Once an interrupt or exception has been recognised, then the processor goes through an internal sequence to switch the processing thread and activate the ISR or exception handler that will service the interrupt or exception. The actual process and, more importantly, the implied work that the service routine must perform varies from processor architecture to architecture. The general processing for an MC68000 or 80x86 which uses a stack frame to hold essential data is different from a RISC processor that uses special internal registers.

Before describing in detail some of the most used mechanisms, let's start with a generic explanation of what is involved. The first part of the sequence is the recognition of the interrupt or exception. This in itself does not necessarily immediately trigger any processor reaction. If the interrupt is not an error condition or the error condition is not connected with the currently executing instruction, the interrupt will not be internally processed until the currently executing instruction has completed. At this point, known as an instruction boundary, the processor will start to internally process the interrupt. If, on the other hand, the interrupt is due to an error with the currently executing instruction, the instruction will be aborted to reach the instruction boundary.

At the instruction boundary, the processor must now save certain state information to allow it to continue its previous execution path prior to the interrupt. This will typically include a copy of the condition code register, the program counter and the return address. This information may be extended to include internal state information as well. The register set is not normally included.

The next phase is to get the location of the ISR to service the interrupt. This is normally kept in a vector table somewhere in memory and the appropriate vector can be supplied by the peripheral or preassigned, or a combination of both approaches. Once the vector has been identified, the processor starts to execute the code within the ISR until it reaches a return from interrupt type of instruction. At this point, the processor, reloads the status information and processing continues the previous instruction stream.

Stack-based processors

With stack-based processors, such as the Intel 80x86, Motorola M68000 family and most 8 bit microcontrollers based on the original microprocessor architectures such as the 8080 and MC6800, the context information that the processor needs to preserve is saved on the external stack.

When the interrupt occurs, the processor context information such as the return address, copies of the internal status registers and so on are stored out on the stack in a stack frame. These stack frames can vary in size and content depending on the source of the interrupt or exception. When the interrupt processing is completed, the information is extracted back from the stack and used to restore the processing prior to the interrupt. It is possible to nest interrupts so that several stack frames and interrupt routines must be executed prior to the program flow being restored. The number of routines that can be nested in this way depends on the storage space available. With external stacks, this depends in turn on the amount of available memory.

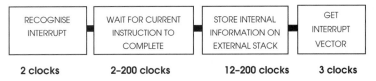

RECOGNISE INTERRUPT	WAIT FOR CURRENT INSTRUCTION TO COMPLETE	STORE INTERNAL INFORMATION ON EXTERNAL STACK	GET INTERRUPT VECTOR
2 clocks	2–200 clocks	12–200 clocks	3 clocks

A typical processor interrupt sequence

Other processors use an internal hardware stack to reduce the external memory cycles necessary to store the stack frame. These hardware stacks are limited in the number of interrupts or exceptions that can be nested. It then falls to the software designer to ensure that this limit is not exceeded.

To show these different interrupt techniques, let's look at some processor examples.

MC68000 interrupts

The MC68000 interrupt and exception processing is based on using an external stack to store the processor's context information. This is very common and similar methods are provided on the 80x86 family and many of the small 8 bit microcontrollers.

Seven interrupt levels are supported and are encoded onto three interrupt pins IP0–IP2. With all three signals high, no external interrupt is requested. With all three asserted, a non-maskable level 7 interrupt is generated. Levels 1–6, generated by other combinations, can be internally masked by writing to the appropriate bits within the status register.

The interrupt cycle is started by a peripheral generating an interrupt. This is usually encoded using a LS148 seven to three priority encoder. This converts seven external pins into a 3 bit binary code. The appropriate code sequence is generated and drives the interrupt pins. The processor samples the levels and requires the levels to remain constant to be recognised. It is recommended that the interrupt level remains asserted until its interrupt acknowledgement cycle commences to ensure recognition.

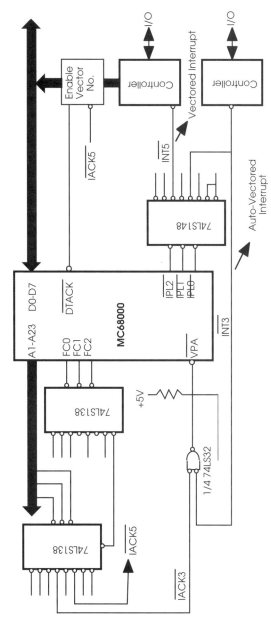

An example MC68000 interrupt design

Once the processor has recognised the interrupt, it waits until the current instruction has been completed and starts an interrupt acknowledgement cycle. This starts an external bus cycle with all three function code pins driven high to indicate an interrupt acknowledgement cycle.

The interrupt level being acknowledged is placed on address bus bits A1–A3 to allow external circuitry to identify which level is being acknowledged. This is essential when one or more interrupt requests are pending. The system now has a choice over which way it will respond:

- If the peripheral can generate an 8 bit vector number, this is placed on the lower byte of the address bus and DTACK* asserted. The vector number is read and the cycle completed. This vector number then selects the address and subsequent software handler from the vector table.

- If the peripheral cannot generate a vector, it can assert VPA* and the processor will terminate the cycle using the M6800 interface. It will select the specific interrupt vector allocated to the specific interrupt level. This method is called auto-vectoring.

To prevent an interrupt request generating multiple acknowledgements, the internal interrupt mask is raised to the interrupt level, effectively masking any further requests. Only if a higher level interrupt occurs will the processor nest its interrupt service routines. The interrupt service routine must clear the interrupt source and thus remove the request before returning to normal execution. If another interrupt is pending from a different source, it can be recognised and cause another acknowledgement to occur.

A typical circuit is shown. Here, level 5 has been allocated as a vectored interrupt and level 3 auto-vectored. The VPA* signal is gated with the level 3 interrupt to allow level 3 to be used with vectored or auto-vectored sources in future designs.

RISC exceptions

RISC architectures have a slightly different approach to exception handling compared to that of CISC architectures. This difference can catch designers out.

Taking the PowerPC architecture as an example, there are many similarities: an exception is still defined as a transition from the user state to the supervisor state in response to either an external request or error, or some internal condition that requires servicing. Generating an exception is the only way to move from the user state to the supervisor state. Example exceptions include external interrupts, page faults, memory protection violations and bus errors. In many ways the exception handling is similar to that used with CISC processors, in that the processor changes to the supervisor state, vectors to an exception handler routine, which investigates the exception and services it before returning control to the original program. This general principle still holds but there are fundamental differences which require careful consideration.

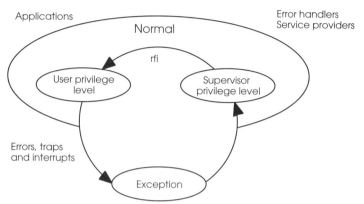

The exception transition model

When an exception is recognised, the address of the instruction to be used by the original program when it restarts and the machine state register (MSR) are stored in the supervisor registers, SRR0 and SRR1. The processor moves into the supervisor state and starts to execute the handler, which resides at the associated vector location in the vector table. The handler can, by examining the DSISR and FPSCR registers, determine the exact cause and rectify the problem or carry out the required function. Once completed, the rfi instruction is executed. This restores the MSR and the instruction address from the SRR0 and SRR1 registers and the interrupted program continues.

There are four general types of exception: asynchronous precise or imprecise and synchronous precise and imprecise. Asynchronous and synchronous refer to when the exception is caused: a synchronous exception is one that is synchronised, i.e. caused by the instruction flow. An asynchronous exception is one where an external event causes the exception; this can effectively occur at any time and is not dependent on the instruction flow. A precise exception is where the cause is precisely defined and is usually recoverable. A memory page fault is a good example of this. An imprecise exception is usually a catastrophic failure, where the processor cannot continue processing or allow a particular program or task to continue. A system reset or memory fault while accessing the vector table falls into this category.

Synchronous precise

All instruction caused exceptions are handled as synchronous precise exceptions. When such an exception is encountered during program execution, the address of either the faulting instruction or the one after it is stored in SRR0. The processor will have completed all the preceding instructions; however, this does not guarantee that all memory accesses caused by these instructions are complete. The faulting instruction will be in an indeterminate state, i.e. it may have started and be partially or completely completed. It is up to the exception handler to determine the instruction type and its completion status using the information bits in the DSISR and FPSCR registers.

Synchronous imprecise

This is generally not supported within the PowerPC architecture and is not present on the MPC601, MPC603 or MCP604 implementations. However, the PowerPC architecture does specify the use of synchronous imprecise handling for certain floating point exceptions and so this category may be implemented in future processor designs.

Asynchronous precise

This exception type is used to handle external interrupts and decrementer-caused exceptions. Both can occur at any time within the instruction processing flow. All instructions being processed before the exceptions are completed, although there is no guarantee that all the memory accesses have completed. SRR0 stores the address of the instruction that would have been executed if no interrupt had occurred.

These exceptions can be masked by clearing the EE bit to zero in the MSR. This forces the exceptions to be latched but not acted on. This bit is automatically cleared to prevent this type of interrupt causing an exception while other exceptions are being processed.

The number of events that can be latched while the EE bit is zero is not stated. This potentially means that interrupts or decrementer exceptions could be missed. If the latch is already full, any subsequent events are ignored. It is therefore recommended that the exception handler performs some form of handshaking to ensure that all interrupts are recognised.

Asynchronous imprecise

Only two types of exception are associated with this: system resets and machine checks. With a system reset all current processing is stopped, all internal registers and memories are reset; the processor executes the reset vector code and effectively restarts processing.

The machine check exception is only taken if the ME bit of the MSR is set. If it is cleared, the processor enters the checkstop state.

Recognising RISC exceptions

Recognising an exception in a superscalar processor, especially one where the instructions are executed out of program order, can be a little tricky — to say the least. The PowerPC architecture handles synchronous exceptions (i.e. those caused by the instruction stream) in strict program order, even though instructions further on in the program flow may have already generated an exception. In such cases, the first exception is handled as if the following instructions have never been executed and the preceding ones have all completed.

There are occasions when several exceptions can occur at the same time. Here, the exceptions are handled on a priority basis using the priority scheme shown in the table below. There is additional priority for synchronous precise exceptions because it is possible for an instruction to generate more than one exception. In these cases, the exceptions would be handled in their own priority order as shown below.

Class	Priority	Description
Async imprecise	1	System reset
	2	Machine check
Sync precise	3	Instruction dependent
Async precise	4	External interrupt
	5	Decrementer interrupt

Exception class priority

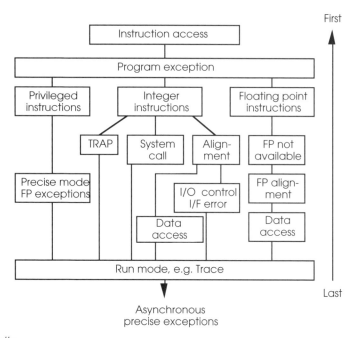

Precise exceptions priority

If, for example, with the single-step trace mode enabled, an integer instruction executed and encountered an alignment error, this exception would be handled before the trace exception. These synchronous precise priorities all have a higher priority than the level 4 and 5 asynchronous precise exceptions, i.e. the external interrupt and decrementer exceptions.

When an exception is recognised, the continuation instruction address is stored in SRR0 and the MSR is stored in SRR1. This saves the machine context and provides the interrupted program with the ability to continue. The continuation instruction may not have started, or be partially or fully complete, depending on the nature of the exception. The FPSCR and DSISR registers contain further diagnostic information for the handler.

When in this state, external interrupts and decrementer exceptions are disabled. The EE bit is cleared automatically to prevent such asynchronous events from unexpectedly causing an exception while the handler is coping with the current one.

It is important to note that the machine status or context which is necessary to allow normal execution to continue is automatically stored in SRR0 and SRR1 — which overwrites the previous contents. As a result, if another exception occurs during an exception handler execution, the result can be catastrophic: the exception handler's machine status information in SRR0 and SRR1 would be overwritten and lost. In addition, the status information in FPSCR and DSISR is also overwritten. Without this information, the handler cannot return to the original program. The new exception handler takes control, processes its exception and, when the rfi instruction is executed, control is passed back to the first exception handler. At this point, this handler does not have its own machine context information to enable it to return control to the original program. As a result the system will, at best, have lost track of that program; at worst, it will probably crash.

This is not the case with the stack-based exception handlers used on CISC processors. With these architectures, the machine status is stored on the stack and, provided there is sufficient stack available, exceptions can safely be nested, with each exception context safely and automatically stored on the stack.

It is for this reason that the EE bit is automatically cleared to disable the external and decrementer interrupts. Their asynchronous nature means that they could occur at any time and if this happened at the beginning of an exception routine, that routine's ability to return control to the original program would be lost. However, this does impose several constraints when programming exception handlers. For the maximum performance in the exception handler, it cannot waste time by saving the machine status information on a stack or elsewhere. In this case, exception handlers should prevent any further exceptions by ensuring that they:

- reside in memory and not be swapped out;
- have adequate stack and memory resources and not cause page faults;
- do not enable external or decrementer interrupts;
- do not cause any memory bus errors.

For exception handlers that require maximum performance but also need the best security and reliability, they should immediately save the machine context, i.e. SRR registers FPSCR and DSISR, preferably on a stack before continuing execution.

In both cases, if the handler has to use or modify any of the user programming model, the registers must be saved prior to modification and they must be restored prior to passing control back. To minimise this process, the supervisor model has access to four additional general-purpose registers which it can use independently of the general-purpose register file in the user programming model.

Enabling RISC exceptions

Some exceptions can be enabled and disabled by the supervisor by programming bits in the MSR. The EE bit controls external interrupts and decrementer exceptions. The FE0 and FE1 bits control which floating point exceptions are taken. Machine check exceptions are controlled via the ME bit.

Returning from RISC exceptions

As mentioned previously, the rfi instruction is used to return from the exception handler to the original program. This instruction synchronises the processor, restores the instruction address and machine state register and the program restarts.

The vector table

Once an exception has been recognised, the program flow changes to the associated exception handler contained in the vector table.

The vector table is a 16 kbyte block (0 to $3FFF) that is split into 256 byte divisions. Each division is allocated to a particular exception or group of exceptions and contains the exception handler routine associated with that exception. Unlike many other architectures, the vector table does not contain pointers to the routines but the actual instruction sequences themselves. If the handler is too large to fit in the division, a branch must be used to jump to its continuation elsewhere in memory.

The table can be relocated by changing the EP bit in the machine state register (MSR). If cleared, the table is located at $0000000. If the bit is set to one (its state after reset) the vector table is relocated to $FFF00000. Obviously, changing this bit before moving the vector table can cause immense problems!

Identifying the cause

Most programmers will experience exception processing when a program has crashed or a cryptic message is returned from a system call. The exception handler can provide a lot of information about what has gone wrong and the likely cause. In this section, each exception vector is described and an indication of the possible causes and remedies given.

The first level investigation is the selection of the appropriate exception handler from the vector table. However, the exception handler must investigate further to find out the exact cause before trying to survive the exception. This is done by checking the information in the FPSCR, DSISR, DAR and MSR registers, which contain different information for each particular vector.

Vector Offset (hex)	Exception	
0 0000	Reserved	
0 0100	System Reset	Power-on, Hard & Soft Resets
0 0200	Machine Check	Eabled through MSR (ME)
0 0300	Data Access	Data Page Fault/Memory Protection
0 0400	Instruction Access	Instr. Page Fault/Memory Protection
0 0500	External Interrupt	INT
0 0600	Alignment	Access crosses Segment or Page
0 0700	Program	Instr. Traps, Errors, Illegal, Privileged
0 0800	Floating-Point Unavailiable	MSR(FP)=0 & F.P. Instruction encountered
0 0900	Decrementer	Decrementer Register passes through 0
0 0A00	Reserved	
0 0B00	Reserved	
0 0C00	System Call	'sc' instruction
0 0D00	Trace	Single-step instruction trace
0 0E00	Floating-Point Assist	A floating-point exception

The basic PowerPC vector table

Fast interrupts

There are other interrupt techniques which greatly simplify the whole process but in doing so provide very fast servicing at the expense of several restrictions. These so-called fast interrupts are often used on DSP processors or microcontrollers where a small software routine is executed without saving the processor context.

This type of support is available on the DSP56000 signal processors, for example. External interrupts normally generate a fast interrupt routine exception. The external interrupt is synchronised with the processor clock for two successive clocks, at which point the processor fetches the two instructions from the appropriate vector location and executes them. Once completed, the program counter simply carries on as if nothing has happened. The advantage is that there is no stack frame building or any other such delays. The disadvantage concerns the size of the routine that can be executed and the resources allocated. When using such a technique, it is usual to allocate a couple of address registers for the fast interrupt routine to use. This allows coefficient tables to be built, almost in parallel with normal execution.

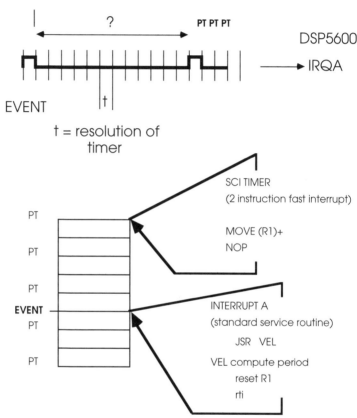

Using interrupts to time an event

The SCI timer is programmed to generate a two instruction fast interrupt which simply auto-increments register R1. This acts as a simple counter which times the period between events. The event itself generates an IRQA interrupt, which forces a standard service routine. The exception handler jumps to routine VEL which processes the data (i.e. takes the value of R and uses it to compute the period), resets R1 and returns from the interrupt.

Interrupt controllers

In many embedded systems there are more external sources for interrupts than interrupt pins on the processor. In this case, it is necessary to use an interrupt controller to provide a larger number of interrupt signals.

An interrupt controller performs several functions:

- It provides a large number of interrupt pins that can be allocated to many external devices. Typically this is at least eight and higher number can be supported by cascading two or more controllers together. This is done on the IBM PC AT where two 8 port controllers are cascaded to give 15 interrupt levels. One level is lost to provide the cascade link.

- It orders the interrupt pins in a priority level so that a high level interrupt will inhibit a lower level interrupt.

- It may provide registers for each interrupt pin which contain the vector number to be used during an acknowledge cycle. This allows peripherals that do not have the ability to provide a vector to do so.

- They can provide interrupt masking. This allows the system software to decide when and if an interrupt is allowed to be serviced. The controller through the use of masking bits within a controller can prevent an interrupt request from being passed through to the processor. In this way, the system has a multi-level approach to screening interrupts. It uses the screening provided by the processor to provide coarse grain granularity while the interrupt controller provides a finer level.

Instruction restart and continuation

The method of continuing the normal execution after exception processing due to a mid-instruction fault, such as caused by a bus error or a page fault, can be done in one of two ways. Instruction restart effectively backs up the machine to the point in the instruction flow where the error occurred. The processor re-executes the instruction and carries on. The instruction continuation stores all the internal data and allows the errant bus cycle to be restarted, even if it is in the middle of an instruction.

The continuation mechanism is undoubtedly easier for software to handle, yet pays the penalty of having extremely large stack frames or the need to store large amounts of context information to allow the processor to continue mid-instruction. The restart mechanism is easier from a hardware perspective, yet can pose increased software overheads. The handler has to determine how far back to restart the machine and must ensure that resources are in the correct state before commencing.

The term 'restart' is important and has some implications. Unlike many CISC processors (for example, the MC68000, MC68020 and MC68030) the instruction does not continue; it is restarted from the beginning. If the exception occurred in the middle of the instruction, the restart repeats the initial action. For many instructions this may not be a problem — but it can lead to some interesting situations concerning memory and I/O accesses.

If the instruction is accessing multiple memory locations and fails after the second access, the first access will be repeated. The store multiple type of instruction is a good example of this, where the contents of several registers are written out to memory. If the target address is an I/O peripheral, an unexpected repeat access may confuse it.

While the majority of the M68000 and 80x86 families are of the continuation type. The MC68040 and PowerPC families along with most microcontrollers — especially those using RISC architectures — are of the restart type. As processors increase in speed and complexity, the penalty of large stack frames shifts the balance in favour of the restart model.

Interrupt latency

One of the most important aspects of using interrupts is in the latency. This is usually defined as the time taken by the processor from recognition of the interrupt to the start of the ISR. It consists of several stages and is dependent on both hardware and software factors. Its importance is that it defines several aspects of an embedded system with reference to its ability to respond to real-time events.

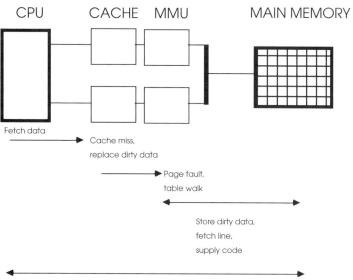

The impact of a cache miss

With a simple microcontroller, this calculation is simple: take the longest execution time for any instruction, add to it the number of memory accesses that the processor needs times the number of clocks per access, add any other delays and you arrive with a worst case interrupt latency. This becomes more difficult when other factors are added such as memory management and caches. While the basic calculations are the same, the number of clocks involved in a memory access can vary dramatically — often by an order of magnitude!

While all RISC systems should be designed with single cycle memory access for optimum performance, the practicalities are that memory cycles often incur wait states or bus delays. Unfortunately, RISC architectures cannot tolerate such delays — one wait state halves the performance, two reduces performance to a third. This can have a dramatic effect on real-time performance. All the advantages gained with the new architecture may be lost.

The solution is to use caches to speed up the memory access and remove delays. This is often used in conjunction with memory management to help control the caches and ensure data coherency, as well as any address translation. However, there are some potential penalties for any system that uses caches and memory management which must be considered.

Consider the system in the diagram. The processor is using a Harvard architecture with combined caches and memory management units to buffer it from the slower main memory. The caches are operating in copyback mode to provide further speed improvements. The processor receives an interrupt and immediately starts exception processing. Although the internal context is preserved in shadow registers, the vector table, exception routines and data exist in external memory.

In this example, the first data fetch causes a cache miss. All the cache lines are full and contain some dirty data, therefore the cache must update main memory with a cache line before fetching the instruction. This involves an address translation, which causes a page fault. The MMU now has to perform an external table walk before the data can be stored. This has to complete before the cache line can be written out which, in turn, must complete before the first instruction of the exception routine can be executed. The effect is staggering — the quick six cycle interrupt latency is totally overshadowed by the 12 or so memory accesses that must be completed simply to get the first instruction. This may be a worst case scenario, but it has to be considered in any real-time design.

This problem can be contained by declaring exception routines and data as non-cachable, or through the use of a BATC or transparent window to remove the page fault and table walk. These techniques couple the CPU directly to external memory which, if slow, can be extremely detrimental to performance. Small areas of very fast memory can be reserved for exception handling to solve this problem; locking routines in cache can also be solutions, at the expense of performance in other system functions.

Do's and Don'ts

This last section describes the major problems that are encountered with interrupt and exceptions, and, more importantly, how to avoid them.

Always expect the unexpected interrupt

Always include a generic handler for all unused/unexpected exceptions. This should record as much information about the processor context such as the exception/vector table number, the return address and so on. This allows unexpected exceptions to be detected and recognised instead of causing the processor and system to crash with little or no hope of finding what has happened.

Control resource sharing

If a resource such as a variable is used by the normal software and within an interrupt routine, care must be taken to prevent corruption.

For example, if you have a piece of C code that modifies a variable a as shown in the example, the expected output would be a=6 if a was 3.

```
{
read(a);
a=2*a;
printf("a=", a);
}
```

If variable a was used in an interrupt routine then there is a risk that the original code will fail, e.g. it would print out a=8, or some other incorrect value. The explanation is that the interrupt routine was executed in the middle of the original code. This changed the value of a and therefore the wrong value was returned.

```
{
read(a);
                    Interrupt!
                    read(a);
                    Return;
a=2*a;
printf("a=", a);
}
```

Exceptions and interrupts can occur asynchronously and therefore if the system shares any resource such as data, or access to peripherals and so on, it is important that any access is handled in such a way that an interrupt cannot corrupt the program flow. This is normally done by masking interrupts before access and unmasking them afterwards. The example code has been modified to include the mask_int and unmask_int calls. The problem is that while the interrupts are masked out, the interrupt latency is higher and therefore this is not a good idea for all applications.

```
{
mask_int();
read(a);
a=2*a;
```

```
printf("a=", a);
unmask_int();
}
```

The best way to solve the problem in the first place is to redesign the software so that the sharing is removed or uses a messaging protocol that copies data to ensure that the corruption cannot take place. This problem is often the cause of obscure the system works fine for days and then crashes type problems where the fault only occurs when certain events happen within certain time frames.

Beware false interrupts

Ensure that all the hardware signals and exception routines do not generate false interrupts. This can happen in software when the interrupt mask or the interrupt handler executes the return from interrupt instruction before the original interrupt source is removed.

In hardware, this can be caused by pulsing the interrupt line and assuming that the processor will only recognise the first pulse and mask out the others. Noise and other factors can corrupt the interrupt lines so that the interrupt is not recognised correctly.

Controlling interrupt levels

This was touched on earlier when controlling resources. It is important to assign high priority events to high priority interrupts. If this does not happen then priority inversion can occur where the lower priority event is serviced while higher priority events wait. This is quite a complex topic and is discussed in more detail in the chapter on real-time operating systems.

Controlling stacks

It is important to prevent stacks from overflowing and exceeding the storage space, whether it is external or internal memory. Some software, in an effort to optimise performance, will remove stack frames from the stack so that the return process can go straight back to the initial program. This is common with nested routines and can be a big time saver. However, it can also be a major source of problems if the frames are not correctly removed or if they are when information must be returned. Another common mistake is to assume that all exceptions have the same size stack frames for all exceptions and all processor models within the family. This is not always the case!

7   Real-time operating systems

What are operating systems?

Operating systems are software environments that provide a buffer between the user and the low level interfaces to the hardware within a system. They provide a constant interface and a set of utilities to enable users to utilise the system quickly and efficiently. They allow software to be moved from one system to another and therefore can make application programs hardware independent. Program debugging tools are usually included which speed up the testing process. Many applications do not require any operating system support at all and run direct on the hardware.

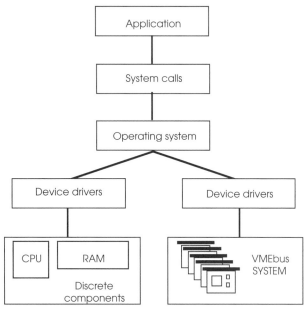

Hardware independence through the use of
an operating system

Such software includes its own I/O routines, for example, to drive serial and parallel ports. However, with the addition of mass storage and the complexities of disk access and file structures, most applications immediately delegate these tasks to an operating system.

The delegation decreases software development time by providing system calls to enable application software access to any of the I/O system facilities. These calls are typically made by building a parameter block, loading a specified register with its location and then executing a software interrupt instruction.

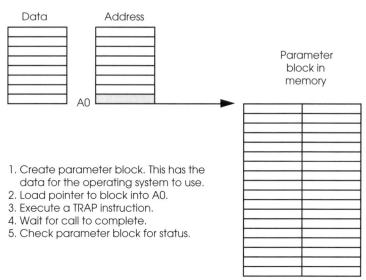

1. Create parameter block. This has the data for the operating system to use.
2. Load pointer to block into A0.
3. Execute a TRAP instruction.
4. Wait for call to complete.
5. Check parameter block for status.

Typical system call mechanism for the M680x0 processor family

The TRAP instruction is the MC68000 family equivalent of the software interrupt and switches the processor into supervisor mode to execute the required function. It effectively provides a communication path between the application and the operating system kernel. The kernel is the heart of the operating system which controls the hardware and deals with interrupts, memory usage, I/O systems etc. It locates a parameter block by using an address pointer stored in a predetermined address register. It takes the commands stored in a parameter block and executes them. In doing so, it is the kernel that drives the hardware, interpreting the commands passed to it through a parameter block. After the command is completed, status information is written back into the parameter block, and the kernel passes control back to the application which continues running in USER mode. The application will find the I/O function completed with the data and status information written into the parameter block. The application has had no direct access to the memory or hardware whatsoever.

These parameter blocks are standard throughout the operating system and are not dependent on the actual hardware performing the physical tasks. It does not matter if the system uses an MC68901 multifunction peripheral or a 8530 serial communication controller to provide the serial ports: the operating system driver software takes care of the dependencies. If the parameter blocks are general enough in their definition, data can be supplied from almost any source within the system, for example a COPY utility could use the same blocks to get data from a serial port and copy it to a parallel port, as copying data from one file to another. This idea of device independence and unified I/O allows software to be reused rather than rewritten. Software can be easily moved from one system to another. This is important for modular embedded designs, especially those that use an industry standard bus such as VMEbus, where system hardware can easily be upgraded and/or expanded.

Operating system internals

The first widely used operating system was CP/M, developed for the Intel 8080 microprocessor and 8" floppy disk systems. It supported I/O calls by two jump tables — BDOS (basic disk operating system) and BIOS (basic I/O system). It quickly became a standard within the industry and a large amount of application software became available for it. Many of the micro-based business machines of the late 1970s and early 1980s were based on CP/M. Its ideas even formed the basis of MSDOS, chosen by IBM for its personal computers.

CP/M is a good example of a single tasking operating system. Only one task or application can be executed at any one time and therefore it only supports one user at a time. When an application is loaded, it provides the user-defined part of the total 'CP/M' program.

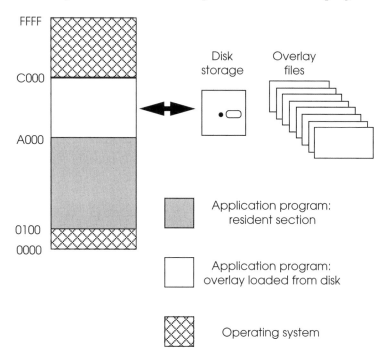

Program overlays

Any application program has to be complete and therefore the available memory often becomes the limiting factor. Program overlays are often used to solve this problem. Parts of the complete program are stored separately on disk and retrieved and loaded over an unused code area when needed. This allows applications larger than the available memory to run, but it places the control responsibility on the application. This is similar to virtual memory schemes where the operating system divides a task's memory into pages and swaps them between memory and mass storage. However, the operating system assumes complete control and such schemes are totally transparent to the user.

With a single tasking operating system, it is not possible to run multiple tasks simultaneously. Large applications have to be run sequentially and cannot support concurrent operations. There is no support for message passing or task control, which would enable applications to be divided into separate entities. If a system needs to take log data and store it on disk and, at the same time, allow a user to process that data using an online database package, a single tasking operating system would need everything to be integrated. With a multitasking operating system, the data logging task can run at the same time as the database. Data can be passed between each element by a common file on disk, and neither task need have any direct knowledge of the other. With a single tasking system, it is likely that the database program would have to be written from scratch. With the multitasking system, a commercially available program can be used, and the logging software interfaced to it. These restrictions forced many applications to interface directly with the hardware and therefore lose the hardware independence that the operating system offered. Such software would need extensive modification to port it to another configuration.

Multitasking operating systems

For the majority of embedded systems, a single tasking operating system is too restrictive. What is required is an operating system that can run multiple applications simultaneously and provide intertask control and communication. The facilities once only available to mini and mainframe computer users are now required by 16/32 bit microprocessor users.

A multitasking operating system works by dividing the processor's time into discrete time slots. Each application or task requires a certain number of time slots to complete its execution. The operating system kernel decides which task can have the next slot, so instead of a task executing continuously until completion, its execution is interleaved with other tasks. This sharing of processor time between tasks gives the illusion to each user that he is the only one using the system.

Context switching, task tables, and kernels

Multitasking operating systems are based around a multitasking kernel which controls the time slicing mechanisms. A time slice is the time period each task has for execution before it is stopped and replaced during a context switch. This is periodically triggered by a hardware interrupt from the system timer. This interrupt may provide the system clock and several interrupts may be executed and counted before a context switch is performed.

When a context switch is performed, the current task is interrupted, the processor's registers are saved in a special table for that particular task and the task is placed back on the 'ready' list to await another time slice. Special tables, often called task control blocks, store all the information the system requires about the task, for example its memory usage, its priority level within the system and its error handling. It is this context information that is switched when one task is replaced by another.

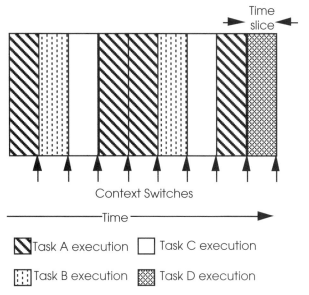

Time slice mechanism for multitasking
operating systems

The 'ready' list contains all the tasks and their status and is used by the scheduler to decide which task is allocated the next time slice. The scheduling algorithm determines the sequence and takes into account a task's priority and present status. If a task is waiting for an I/O call to complete, it will be held in limbo until the call is complete.

Once a task is selected, the processor registers and status at the time of its last context switch are loaded back into the processor and the processor is started. The new task carries on as if nothing had happened until the next context switch takes place. This is the basic method behind all multitasking operating systems.

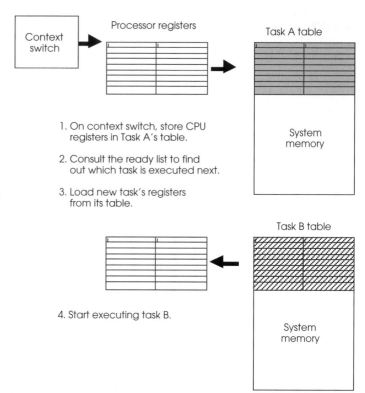

1. On context switch, store CPU registers in Task A's table.

2. Consult the ready list to find out which task is executed next.

3. Load new task's registers from its table.

4. Start executing task B.

Context switch mechanism

The diagram shows a simplified state diagram for a typical real-time operating system which uses this time slice mechanism. On each context switch, a task is selected by the kernel's scheduler from the 'ready' list and is put into the run state. It is then executed until another context switch occurs. This is normally signalled by a periodic interrupt from a timer. In such cases the task is simply switched out and put back on the 'ready' list, awaiting its next slot. Alternatively, the execution can be stopped by the task executing certain kernel commands. It could suspend itself, where it remains present in the system but no further execution occurs. It could become dormant, awaiting a start command from another task, or even simply waiting for a server task within the operating system to perform a special function for it. A typical example of a server task is a driver performing special screen graphics functions. The most common reason for a task to come out of the run state, is to wait for a message or command, or delay itself for a certain time period. The various wait directives allow tasks to synchronise and control each other within the system. This state diagram is typical of many real-time operating systems.

The kernel controls memory usage and prevents tasks from corrupting each other. If required, it also controls memory sharing between tasks, allowing them to share common program modules, such as high level language runtime libraries. A set of memory tables is maintained, which is used to decide if a request is accepted or rejected. This means that resources, such as physical memory and peripheral devices, can be protected from users without using hardware memory management provided the task is disciplined enough to use the operating system and not access the resources directly. This is essential to maintain the system's integrity.

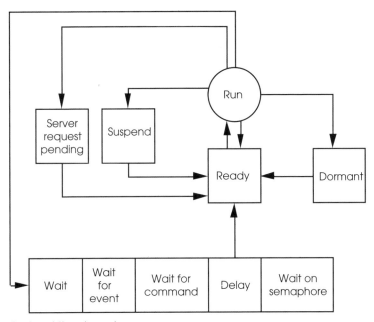

State diagram for a typical real-time kernel

Message passing and control can be implemented in such systems by using the kernel to act as a message passer and controller between tasks. If task A wants to stop task B, then by executing a call to the kernel, the status of task B can be changed and its execution halted. Alternatively, task B can be delayed for a set time period or forced to wait for a message.

With a typical real-time operating system, there are two basic type of messages that the kernel will deal with:

• flags that can control but cannot carry any implicit information — often called semaphores or events

• messages which can carry information and control tasks — often called messages or events.

The kernel maintains the tables required to store this information and is responsible for ensuring that tasks are controlled and receive the information. With the facility for tasks to communicate between each other, system call support for accessing I/O, loading tasks from disk etc., can be achieved by running additional tasks, with a special system status. These system tasks provide additional facilities and can be included as required.

To turn a real-time kernel into a full operating system with file systems and so on, requires the addition of several such tasks to perform I/O services, file handling and file management services, task loading, user interface and driver software. What was about a small <16 kbyte-sized kernel will often grow into a large 120 kbyte operating system. These extra facilities are built up as layers surrounding the kernel. Application tasks then fit around the outside. A typical onion structure is shown as an example. Due to the modular construction, applications can generally access any level directly if required. Therefore, application tasks that just require services provided by the kernel can be developed and debugged under the full environment, and stripped down for integration onto the target hardware.

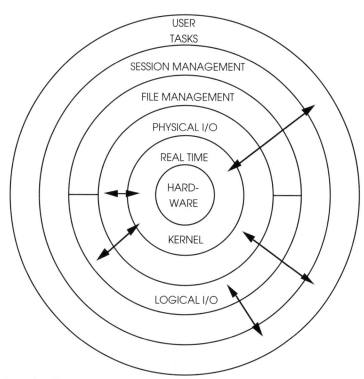

A typical operating system structure

In a typical system, all these service tasks and applications are controlled, scheduled and executed by the kernel. If an application wishes to write some data to a hard disk in the system, the process starts with the application creating a parameter block and asking the file manager to open the file. This system call is normally executed by a TRAP instruction. The kernel then places the task on its 'waiting' list until the file manager had finished and passed the status information back to the application task. Once this event has been received, it wakes up and is placed on the 'ready' list awaiting a time slot.

These actions are performed by the kernel. The next application command requests the file handling services to assign an identifier — often called a logical unit number (LUN) — to the file prior to the actual access. This is needed later for the I/O services call. Again, another parameter block is created and the file handler is requested to assign the LUN. The calling task is placed on the 'waiting' list until this request is completed and the LUN returned by the file handler. The LUN identifies a particular I/O resource such as a serial port or a file without actually knowing its physical characteristics. The device is therefore described as logical rather than physical.

With the LUN, the task can create another parameter block, containing the data, and ask the I/O services to write the data to the file. This may require the I/O services to make system calls of its own. It may need to call the file services for more data or to pass further information on. The data is then supplied to the device driver which actually executes the instructions to physically write the data to the disk. It is generally at this level that the logical nature of the I/O request is translated into the physical characteristics associated with the hardware. This translation should lie in the domain of the device driver software. The user application is unaware of these characteristics.

A complex system call can cause many calls between the system tasks. A program loader that is requested by an application task to load another task from memory needs to call the file services and I/O services to obtain the file from disk, and the kernel to allocate memory for the task to be physically loaded.

The technique of using standard names, files, and/or logical unit numbers to access system I/O makes the porting of application software from one system to another very easy. Such accesses are independent of the hardware the system is running on, and allow applications to treat data received or sent in the same way, irrespective of its source.

What is a real-time operating system?

Many multitasking operating systems available today are also described as 'real-time'. These operating systems provide additional facilities allowing applications that would normally interface directly with the microprocessor architecture to use interrupts and drive peripherals to do so without the operating system blocking such activities. Many multitasking operating systems prevent the user from accessing such sensitive resources. This overzealous caring can prevent many operating systems from being used in applications such as industrial control.

A characteristic of a real-time operating system is its defined response time to external stimuli. If a peripheral generates an interrupt, a real-time system will acknowledge and start to service it within a maximum defined time. Such response times vary from system to system, but the maximum time specified is a worst case figure, and will not be exceeded due to changes in factors such as system workload.

Any system meeting this requirement can be described as real-time, irrespective of the actual value, but typical industry accepted figures for context switches and interrupt response times are about 10 microseconds. This figure gets smaller as processors become more powerful and run at higher speeds. With several processors having the same context switch mechanism, the final context switch time come down to its clock speed and the memory access time.

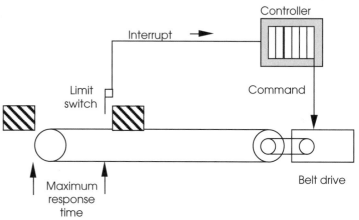

Example of a real-time response

The consequences to industrial control of not having a real-time characteristic can be disastrous. If a system is controlling an automatic assembly line, and does not respond in time to a request from a conveyor belt limit switch to stop the belt, the results are easy to imagine. The response does not need to be instantaneous — if the limit switch is set so that there are 3 seconds to stop the belt, any system with a guaranteed worst case response of less than 3 seconds can meet this real-time requirement.

For an operating system to be real-time, its internal mechanisms need to show real-time characteristics so that the internal processes sequentially respond to external interrupts in guaranteed times.

When an interrupt is generated, the current task is interrupted to allow the kernel to acknowledge the interrupt and obtain the vector number that it needs to determine how to handle it. A typical technique is to use the kernel's interrupt handler to update a linked list which contains information on all the tasks that need to be notified of the interrupt.

If a task is attached to a vector used by the operating system, the system actions its own requirements prior to any further response by the task. The handler then sends an event message to the tasks attached to the vector, which may change their status and completely change the priorities of the task ready list. The scheduler analyses the list, and dispatches the highest priority task to run. If the interrupt and task priorities are high enough, this may be the next time slice.

The diagram depicts such a mechanism: the interrupt handler and linked list searches are performed by the kernel. The first priority is to service the interrupt. This may be from a disk controller indicating that it has completed a data transfer. Once the kernel has satisfied its own needs, the handler will start a linked list search. The list comprises of blocks of data identifying tasks that have their own service routines. Each block will contain a reference to the next block, hence the linked list terminology.

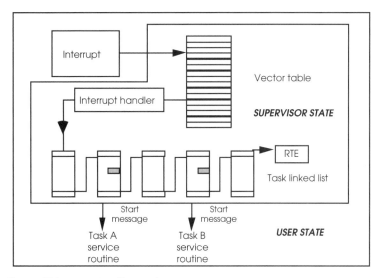

Handling interrupt routines within an operating system

Each identified task is then sent a special message. This will start the task's service routine when it receives its next time slice. The kernel interrupt handler will finally execute an RTE return from exception instruction which will restore the processor state prior to the interrupt. In such arrangements the task service routines execute in USER mode. The only SUPERVISOR operation is that of the kernel and its own interrupt handler. As can be imagined, this processing can increase the interrupt latency seen by the task quite dramatically. A tenfold increase is not uncommon.

To be practical, a real-time operating system has to guarantee maximum response times for its interrupt handler, event passing mechanisms, scheduler algorithm and provide system calls to allow tasks to attach and handle interrupts.

With the conveyor belt example above, a typical software configuration would dedicate a task to controlling the conveyor belt. This task would make several system calls on start-up to access the parallel I/O peripheral that interfaces the system to components such as the drive motors and limit switches and tells the kernel that certain interrupt vectors are attached to the task and are handled by its own interrupt handling routine.

Once the task has set everything up, it remains dormant until an event is sent by other tasks to switch the belt on or off. If a limit switch is triggered, it sets off an interrupt which forces the kernel to handle it. The currently executing task stops, the kernel handler searches the task interrupt attachment linked list, and places the controller task on the ready list, with its own handler ready to execute. At the appropriate time slice, the handler runs, accesses the peripheral and switches off the belt. This result may not be normal, and so the task also sends event messages to the others, informing them that it has acted independently and may force other actions. Once this has been done, the task goes back to its dormant state awaiting further commands.

Real-time operating systems have other advantages: to prevent a system from power failure usually needs a guaranteed response time so that the short time between the recognition of and the actual power failure can be used to store vital data and bring the system down in a controlled manner. Many operating systems actually have a power fail module built into the kernel so that no time is lost in executing the module code.

So far in this chapter, an overview of the basics behind a real-time operating system have been explained. There are, however, several variants available for the key functions such as task swapping and so on. The next few sections will delve deeper into these topics.

Task swapping methods

The choice of scheduler algorithms can play an important part in the design of an embedded system and can dramatically affect the underlying design of the software. There are many different types of scheduler algorithm that can be used, each with either different characteristics or different approaches to solving the same problem of how to assign priorities to schedule tasks so that correct operation is assured.

Time slice

Time slicing has been previously mentioned in this chapter under the topic of multitasking and can be used within an embedded system where time critical operations are not essential. To be more accurate about its definition, it describes the task switching mechanism and not the algorithm behind it although its meaning has become synonymous with both.

Time slicing works by making the task switching regular periodic points in time. This means that any task that needs to run next will have to wait until the current time slice is completed or until the current task suspends its operation. This technique can also be used as a scheduling method as will be explained later in this chapter.

The choice of which task to run next is determined by the scheduling algorithm and thus is nothing to do with the time slice mechanism itself. It just happens that many time slice-based systems use a round-robin or other fairness scheduler to distribute the time slices across all the tasks that need to run.

For real-time systems where speed is of the essence, the time slice period plus the context switch time of the processor determines the context switch time of the system. With most time slice periods in the order of milliseconds, it is the dominant factor in the system response. While the time period can be reduced to improve the system context switch time, it will increase the number of task switches that will occur and this will reduce the efficiency of the system. The larger the number of switches, the less time there is available for processing.

Pre-emption

The alternative to time slicing is to use pre-emption where a currently running task can be stopped and switched out — pre-empted — by a higher priority *active* task. The *active* qualifier is important as the example of pre-emption later in this section will show. The main difference is that the task switch does not need to wait for the end of a time slice and therefore the system context switch is now the same as the processor context switch.

As an example of how pre-emption works, consider a system with two tasks A and B. A is a high priority task that acts as an ISR to service a peripheral and is activated by a processor interrupt from the peripheral. While it is not servicing the peripheral, the task remains dormant and stays in a suspended state. Task B is a low priority task that performs system housekeeping.

When the interrupt is recognised by the processor, the operating system will process it and activate task A. This task with its higher priority compared to task B will cause task B to be pre-empted and replaced by task A. Task A will continue processing until it has completed and then suspend itself. At this point, task B will context switch task A out because task A is no longer active.

This can be done with a time slice mechanism provided the interrupt rate is less than the time slice rate. If it is higher, this can also be fine provided there is sufficient buffering available to store data without losing it while waiting for the next time slice point. The problem comes when the interrupt rate is higher or if there are multiple interrupts and associated tasks. In this case, multiple tasks may compete for the same time slice point and the ability to run even though the total processing time needed to run all of them may be considerably less than the time provided within a single time slot. This can be solved by artificially creating more context switch points by getting each task to suspend after completion. This may offer only a partial solution because a higher priority task may still have to wait on a lower priority task to complete. With time slicing, the lower priority task cannot be pre-empted and therefore the higher priority task must wait for the end of the time slice or the lower priority task to complete. This is a form of priority inversion which is explained in more detail later.

Most real-time operating systems support pre-emption in preference to time slicing although some can support both methodologies

Co-operative multitasking

This is the mechanism behind Windows 3.1 and while not applicable to real-time operating systems for reasons which will become apparent, it has been included for reference.

The idea of co-operative multitasking is that the tasks themselves co-operate between themselves to provide the illusion of multitasking. This is done by periodically allowing other tasks or applications the opportunity to execute. This requires programming within the application and the system can be destroyed by a single rogue program that hogs all the processing power. This method may be acceptable for a desktop personal computer but it is not reliable enough for most real-time embedded systems.

Scheduler algorithms

So far in this section, the main methods of swapping tasks has been discussed. It is clear that pre-emption is the first choice for embedded systems because of its better system response. The next issue to address is how to assign the task priorities so that the system works and this is the topic that is examined now.

Rate monotonic

Rate monotonic scheduling (RMS) is an approach that is used to assign task priority for a pre-emptive system in such a way that the correct execution can be guaranteed. It assumes that task priorities are fixed for a given set of tasks and are not dynamically changed during execution. It assumes that there are sufficient task priority levels for the task set and that the task set models periodic events only. This means that an interrupt that is generated by a serial port peripheral is modelled as an event that occurs on a periodic rate determined by the data rate, for example. Asynchronous events such as a user pressing a key are handled differently as will be explained later.

The key policy within RMS is that tasks with shorter execution periods are given the highest priority within the system. This means that the faster executing tasks can pre-empt the slower periodic tasks so that they can meet their deadlines. The advantage this gives the system designer is that it is easier to theoretically specify the system so that the tasks will meet their deadlines without overloading the processor. This requires detailed knowledge about each task and the time it takes to execute. This and its periodicity can be used to calculate the processor loading.

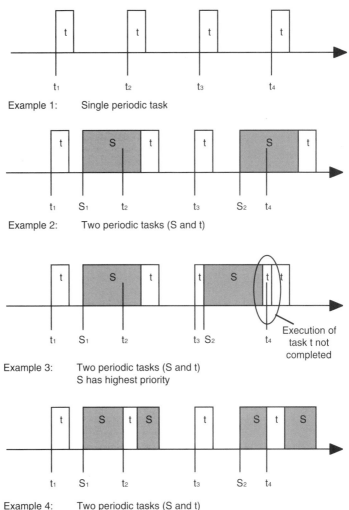

Example 1:   Single periodic task

Example 2:   Two periodic tasks (S and t)

Example 3:   Two periodic tasks (S and t)
               S has highest priority

Example 4:   Two periodic tasks (S and t)
               t has highest priority

Using RMS policies

To see how this policy works, consider the examples shown in the above diagram. In the diagrams, events that start a task are shown as lines that cross the horizontal time line and tasks are shown as rectangles whose length determines their execution time. Example 1 shows a single periodic task where the task t is executed with a periodicity of time t. The second example adds a second task S where its periodicity is longer than that of task t. The task priority shown is with task S having the highest priority. In this case, the RMS policy has not been followed because the longest task has been given a higher priority than the shortest task. However, please note that in this case the system works fine because of the timing of the tasks' periods.

Example 3 shows what can go wrong if the timing is changed and the periodicity for task S approaches that of task t. When t3 occurs, task t is activated and starts to run. It does not complete because S2 occurs and task S is swapped-in due to its higher priority. When tasks S completes, task t resumes but during its execution, the event t4 occurs and thus task t has failed to meets its task 3 deadline. This could result in missed or corrupted data, for example. When task t completes, it is then reactivated to cope with the t4 event.

Example 4 shows the same scenario with the task priorities reversed so that task t pre-empts task S. In this case, RMS policy has been followed and the system works fine with both tasks meeting their deadlines. This system is useful to understand and does allow theoretical analysis before implementation to prevent a design from having to manually assign task priorities to get it to work using a trial and error approach. It is important to remember within any calculations to take into account the context swapping time needed to pre-empt and resume tasks. The processor utilisation can also be calculated and thus give some idea of the performance needed or how much performance is spare. Typically, the higher the utilisation (>80%), the more chance that the priority assignment is wrong and that the system will fail. If the utilisation is below this, the chances are that the system will run correctly.

This is one of the problems with theoretical analysis of such systems in that the actual design may have to break some of the assumptions on which the analysis is based. Most embedded systems will have asynchronous events and tasks running. While these can be modelled as a periodic task that polls for the asynchronous event and are analysed as if they will run, other factors such as cache memory hit ratios can invalidate by lengthening the execution time of any analysis. Similarly, the act of synchronising tasks by inter-task communication can also cause difficulties within any analysis.

Deadline monotonic scheduling

Deadline monotonic scheduling (DMS) is another task priority policy that uses the nearest deadline as the criterion for assigning task priority. Given a set of tasks, the one with the nearest deadline is given the highest priority. This means that the scheduling or designer must now know when these deadlines are to take place. Tracking and, in fact, getting this information in the first place can be difficult and this is often the reason behind why deadline scheduling is often a second choice compared to RMS.

Priority guidelines

With a system that has a large number of tasks or one that has a small number of priority levels, the general rule is to assign tasks with a similar period to the same level. In most cases, this does not effect the ability to schedule correctly.

If a task has a large context, i.e. it has more registers and data to be stored compared to other tasks, it is worth raising its priority to reduce the context switch overhead. This may prevent the system from scheduling properly but can be a worthwhile experiment.

Priority inversion

It is also possible to get a condition called priority inversion where a lower priority task can continue running despite there being a higher priority task active and waiting to pre-empt.

This can occur if the higher priority task is in someway blocked by the lower priority task through it having to wait until a message or semaphore is processed. This can happen for several reasons.

Disabling interrupts

While in the interrupt service routine, all other interrupts are disabled until the routine has completed. This can cause a problem if another interrupt is received and held pending. What happens is that the higher priority interrupt is held pending in favour of the lower priority one — albeit that it occurred first. As a result, priority inversion takes place until interrupts are re-enabled at which point the higher priority interrupt will start its exception processing, thus ending the priority inversion.

One metric of an operating system is the longest period of time that all interrupts are disabled. This must then be added to any other interrupt latency calculation to determine the actual latency period.

Message queues

If a message is sent to the operating system to activate a task, many systems will process the message and then reschedule accordingly. In this way, the message queue order can now define the task priority. For example, consider an ISR that sends an unblocking message to two tasks, A and B, that are blocked waiting for the message. The ISR sends the message for task A first followed by task B. The ISR is part of the RTOS kernel and therefore may be subject to several possible conditions:

• Condition 1

Although the message calls may be completed, their action may be held pending by the RTOS so that any resulting pre-emption is stopped from switching out the ISR.

• Condition 2

The ISR may only be allowed to execute a single RTOS call and in doing so the operating system itself will clean up any stack frames. The operating system will then send messages to tasks notifying them of the interrupt and in this way simulate the interrupt signal. This is normally done through a list.

These conditions can cause priority inversion to take place. With condition 1, the ISR messages are held pending and processed. The problem arises with the methodology used by the operating system to process the pending messages. If it processes all the messages, effectively unblocking both tasks before instigating the scheduler to decide the next task to run, all is well. Task B will be scheduled ahead of task A because of its higher priority. The downside is the delay in processing all the messages before selecting the next task.

Most operating systems, however, only have a single call to process and therefore in normal operation do not expect to handle multiple messages. In this case, the messages are handled individually so that after the first message is processed, task A would be unblocked and allowed to execute. The message for task B would either be ignored or processed as part of the housekeeping at the next context switch. This is where priority inversion would occur. The ISR has according to its code unblocked both tasks and thus would expect the higher priority task B to execute. In practice, only task A is unblocked and is running, despite it being at a lower priority. This scenario is a programming error but one that is easy to make.

To get around this issue, some RTOS implementations restrict an ISR to making either one or no system calls. With no system calls, the operating system itself will treat the ISR event as an internal message and will unblock any task that is waiting for an ISR event. With a single system call, a task would take the responsibility for controlling the message order to ensure that priority inversion does not take place.

Waiting for a resource

If a resource is shared with a low priority task and it does not release it, a higher priority task that needs it can be blocked until it is released. A good example of this is where the interrupt handler is distributed across a system and needs to access a common bus to handle the interrupt. This can be the case with a VMEbus system, for example.

VMEbus interrupt messages

VMEbus is an interconnection bus that was developed in the early 1980s for use within industrial control and other real-time applications. The bus is asynchronous and is very similar to that of the MC68000. It comprises of a separate address, data, interrupt and control buses.

If a VMEbus MASTER wishes to inform another that a message is waiting or urgent action is required, a VMEbus interrupt can be generated. The VMEbus supports seven interrupt priority levels to allow prioritisation of a resource.

Any board can generate an interrupt by asserting one of the levels. Interrupt handling can either be centralised, and handled by one MASTER, or can be distributed among many. For multiprocessor applications, distributed handling allows rapid direct communication to individual MASTERs by any board in the system capable of generating an interrupt: the MASTER that has been assigned to handle the interrupt requests the bus and starts the interrupt acknowledgement cycle. Here, careful consideration of the arbitration level chosen for the MASTER is required. The interrupt response time depends on the time taken by the handler to obtain the bus prior to the acknowledgement. If it has a low priority, the overall response time may be more than that obtained for a lower priority interrupt whose handler has a higher arbitration level. The diagrams below show the relationship for both priority and round robin arbitration schemes. Again, as with the case with arbitration schemes, the round robin system has been assumed on average to provide equal access for all the priority levels.

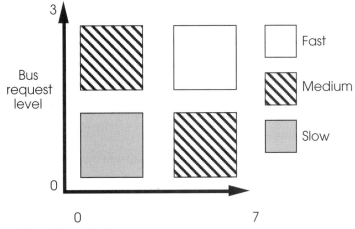

VMEbus interrupt response times for a priority arbitration scheme

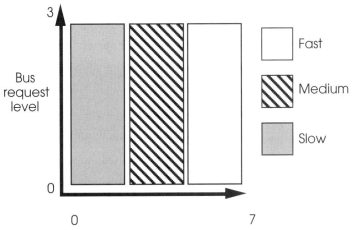

VMEbus interrupt response times for a round robin arbitration scheme

Priority inversion can occur if a lower priority interrupt may be handled in preference to a higher priority one, simply because of the arbitration levels. To obtain the best response, high priority interrupts should only be assigned to high arbitration level MASTERs. The same factors, such as local traffic on the VMEbus and access time, increase the response time as the priority level decreases.

VMEbus only allows a maximum of seven separate interrupt levels and this limits the maximum number of interrupt handlers to seven. For systems with larger numbers of MASTERs, polling needs to be used for groups of MASTERs assigned to a single interrupt level. Both centralised and distributed interrupt handling schemes have their place within a multiprocessor system. Distributed schemes allow interrupts to be used to pass high priority messages to handlers, giving a fast response time which might be critical for a real-time application. For simpler designs, where there is a dominant controlling MASTER, one handler may be sufficient.

Fairness systems

There are times when the system requires different characteristics from those that originally provided or alternatively wants a system response that is not priority based. This can be achieved by using a fairness system where the bus access is distributed across the requesting processors. There are various schemes that are available such as round robin where access is simply passed from one processor to another. Other methods can use time sharing where the bus is given to a processor on the understanding that it must be relinquished within a maximum time frame.

These type of systems can affect the interrupt response because bus access is often necessary to service an interrupt.

Tasks, threads and processes

This section of the chapter discusses the nomenclature given to the software modules running within the RTOS environment. Typically, they have been referred to as tasks but other names such as threads and processes are also used to refer to entities within the RTOS. They are sometimes used instead as an interchangeable replacement for the term task. In practice, they refer to different aspects within the system.

So far in this chapter, a task has been used to describe an entity of work within an operating system that has control over resources. When a context switch is performed, it effectively switches to another task which takes over. Strictly speaking the context switch may include additional information that is relevant to the task such as memory management information, which is beyond the simple register swapping that is performed within the processor. As a result, the term process is often used to encompass more than a simple context switch and thus includes the additional information. The problem is that this is very similar to that of a task switch or context switch that the definitions have become blurred and almost interchangeable.

A task or process has several characteristics:

- It owns or controls resources, e.g. access to peripherals and so on.

- It has threads of execution. These are paths through the code contained within the task or process. Normally, there is a single thread though but this may not always be the case. Multiple threads can be supported if the task or process can maintain separate data areas for each thread. This also requires the code to be written in a re-entrant manner.

- It requires additional information beyond the normal register contents to maintain its integrity, e.g. memory management information, cache flushing and so on. When a new process or task is swapped-in, not only are the processor registers changed but additional work must be done such as invalidating caches to ensure that the new process or task does not access incorrect information.

A thread has different characteristics:

- It has no additional context information beyond that stored in the processor register set.

- Its ownership of resources is inherited from its parent task or process.

With a simple operating system, there is no difference between the thread context switch and the process level switch. As a result, these terms almost become interchangeable. With a multi-user, multitasking operating system, this is not the case. The process or task is the higher level with the thread(s) the lower level.

Some operating systems take this a stage further and define a three level hierarchy: a process consists of a set of tasks with each task having multiple threads. Be warned! These terms mean many different things depending on who is using them.

Exceptions

With most embedded systems, access to the low level exception handler is essential to allow custom routines to be written to support the system. This can include interrupt routines to control external peripherals, emulation routines to simulate instructions or facilities that the processor does not support — software floating point is a very good example of this — and other exception types.

Some of these exceptions are needed by the RTOS to provide entry points into the kernel and to allow the timers and other facilities to function. As a result, most RTOSs already provide the basic functionality for servicing exceptions and provide access points into this functionality to allow the designer to add custom exception routines. This can be done in several ways:

- Patching the vector table

  This is relatively straight forward if the vector is not used by the RTOS. If it is, then patching will still work but the inserted user exception routine must preserve the exception context and then jump to the existing handler instead of using a return from exception type instruction to restore normal processing. If it is sharing an exception with the RTOS, there must be some form of checking so that the user handler does not prevent the RTOS routine from working correctly.

- Adding user orientated routines to existing exception handlers

  This is very similar to the previous technique in that the user routine is added to any existing RTOS routine. The difference is that the mechanism is more formal and does not require vector table patching or any particular checking by the user exception handler.

- Generating a pseudo exception that is handled by separate user exception handler(s)

  This is even more formal — and slower — and effectively replaces the processor level exception routine with a RTOS level version in which the user creates his own vector table and exception routines. Typically, all this is performed through special kernel calls which register a task as the handler for a particular exception. On completion, the handler uses a special return from exception call into the RTOS kernel to signify that it has completed.

Memory model

The memory model that the processor offers can and often varies with the model defined by the operating system and open to the software designer to use. In other words, although the processor may support a full 32 bit address range with full memory mapped I/O and read/write access anywhere in the map at a level of an individual word or map, the operating system's representation of the memory map may only be 28 bits, with I/O access allocated on a 512 byte basis with read only access for the first 4 Mbytes of RAM and so on.

This discrepancy can get even wider, the further down in the levels that you go. For example, most processors that have sophisticated cache memory support use the memory management unit. This then requires fairly detailed information about the individual memory blocks within a system. This information has to be provided to the RTOS and is normally done using various memory allocation techniques where information is provided when the system software is compiled and during operation.

Memory allocation

Most real-time operating systems for processors where the memory map can be configured, e.g. those that have large memory addressing and use memory mapped I/O, get around this problem by using a special file that defines the memory map that the system is expected to use and support. This will define which memory addresses correspond to I/O areas, RAM, ROM and so on. When a task is created, it will be given a certain amount of memory to hold its code area and provide some initial data storage. If it requires more memory, it will request it from the RTOS using a special call such as malloc(). The RTOS will look at the memory request and allocate memory by passing back a pointer to the additional memory. The memory request will normally define the memory characteristics such as read/write access, size, and even its location and attributes such as physical or logical addressing.

The main question that arises is why dynamically allocate memory? Surely this can be done when the tasks are built and included with the operating system? The answer is not straightforward. In simple terms, it is a yes and for many simple embedded systems, this is correct. For more complex systems, however, this static allocation of memory is not efficient, in that memory may be reserved or allocated to a task yet could only be used rarely within the system's operation. By dynamically allocating memory, the total amount of memory that the system requires for a given function can can be reduced by reclaiming and reallocation memory as and when needed by the system software. This will be explained in more detail later on.

Memory characteristics

The memory characteristics are important to understand especially when different memory addresses correspond to different physical memory. As a result, asking for a specific block of memory may impact the system performance. For example, consider an embedded processor that has both internal and external memory. The internal memory is faster than the external memory and therefore improves performance by not inserting wait states during a memory access. If a task asks for memory expecting to be allocated internal memory but instead receives a pointer to external memory, the task performance will be degraded and potentially the system can fail. This is not a programming error in the true sense of the word because the request code and RTOS have executed correctly. If the request was not specific enough, then the receiving task should expect the worst case type of memory. If it does not or needs specific memory, this should be specified during the request. This is usually done by specifying a memory address or type that the RTOS memory allocation code can check against the memory map file that was used when the system was built.

- Read/write access

  This is straightforward and defines the access permissions that a task needs to access a memory block.

- Internal/external memory

  This is normally concerned with speed and performance issues. The different types of memory are normally defined not by their speed but indirectly through the address location. As a result, the programmer must define and use a memory map so that the addresses of the required memory block match up the required physical memory and thus its speed. Some RTOSs actually provide simple support flags such as internal/external but this is not common.

- Size

  The minimum and maximum sizes are system dependent and typically are influenced by the page size of any memory management hardware that may be present. Some systems can return with partial blocks, e.g. if the original request was for 8 kbytes, the RTOS may only have 4 kbytes free and instead of returning an error, will return a pointer to the 4 kbytes block instead. This assumes that the requesting task will check the returned size and not simply assume that because there was no error, it has all the 8 kbytes it requested! Check the RTOS details carefully.

- I/O

  This has several implications when using processors that execute instructions out of order to remove pipeline stalls and thus gain performance. Executing instructions that access I/O ports out of sequence can break the program syntax and integrity. The program might output a byte and then read a status register. If this is reversed, the correct sequence has been destroyed and the software will probably crash. By declaring I/O addresses as I/O, the processor can be programmed to ensure the correct sequence whenever these addresses are accessed.

- Cached or non-cachable

  This is similar to the previous paragraph on I/O. I/O addresses should not be cached to prevent data corruption. Shared memory blocks need careful handling with caches and in many cases unless there is some form of bus snooping to check that the contents of a cache is still valid, these areas should also not be cached.

- Coherency policies

  Data caches can have differing coherency policies such as write through, copy back and so on which are used to ensure the data coherency within the system. Again, the ability to specify or change these policies is useful.

Example memory maps

The first example is that commonly used within a simple microcontroller where its address space is split into the different memory types. The example shows three: I/O devices and peripherals, program RAM and ROM and data RAM. The last two types have then been expanded to show how they could be allocated to a simple embedded system. The program area contains the code for four tasks, the RTOS code and the processor vector table. The data RAM is split into five areas: one for each of the tasks and a fifth area for the stack. In practice, these areas are often further divided into internal and external memory, EPROM and EEPROM, SRAM and even DRAM, depending on the processor architecture and model. This example uses a fixed static memory map where the memory requirements for the whole system are defined at compile and build time. This means that tasks cannot get access to additional memory by using some of the memory allocated to another task. In addition, it should be remembered that although the memory map shows nicely partitioned areas, it does not imply nor should it be assumed that task A cannot access task C's data area, for example. In these simple processors and memory maps, all tasks have the ability to access any memory location and it only the correct design and programming that ensures that there is no corruption. Hardware can be used to provide this level of protection but it requires some form of memory management unit to check that programs are conforming to their design and not accessing memory that they should not. Memory management is explained in some detail in the next section.

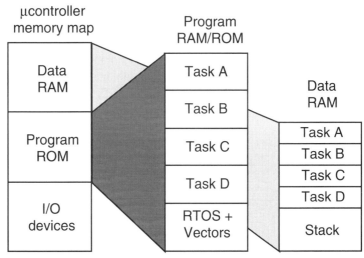

Simple microcontroller memory map

The second example shows a similar system to the first example except that it has been further partitioned into internal and external memory. The internal memory runs faster than the external memory and because it does not have any wait states, its access time is faster and the processor performance does not degrade. The slower external memory has two wait states and with a single cycle processor would degrade performance by 66% — each instruction would take three clocks instead of one, for example.

Given this performance difference, it is important that the memory resources are carefully allocated. In the example, task A requires the best performance and the system needs fast task switching. This means that both the task A code and data, along with the RTOS and the system stack, are allocated to the internal memory where it will get the best performance. All other task code and data are stored externally because all the internal memory is used.

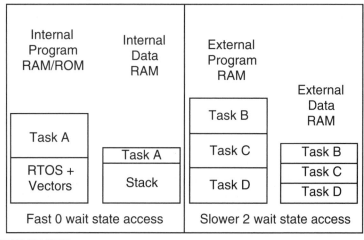

Internal and external memory map

The third example shows a dynamic allocation system where tasks can request additional memory as and when they need it. The first map shows the initial state with the basic memory allocated to tasks and RTOS. This is similar to the previous examples except that there is a large amount of memory entitled dynamic memory which is controlled by the RTOS and can be allocated dynamically by it to other tasks on demand. The next two diagrams show this in operation.

The first request by task C starts by sending a request to the RTOS for more memory. The RTOS allocates a block to the task and returns a pointer to it. Task C can use this to get access to this memory. This can be repeated if needed and the next diagram shows task C repeating the request and getting access to a second block. Blocks can also be relinquished and returned to the RTOS for allocation to other tasks at some other date.

This process is highly dynamic and thus can provide a mechanism for minimising memory usage. Task C could be allocated all the memory at certain times and as the memory is used and no longer required, blocks can be reallocated to different tasks.

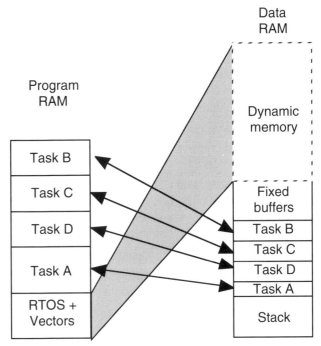

Memory map with dynamic allocation — initial state

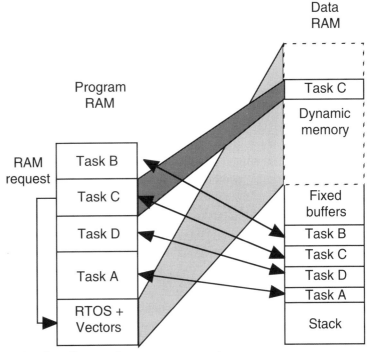

Memory map with dynamic allocation — after memory request

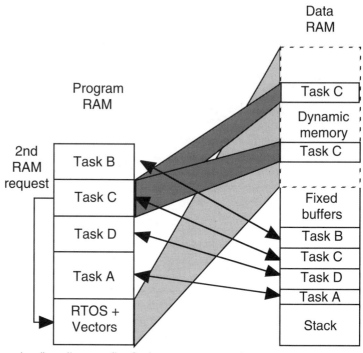

Memory map with dynamic allocation — after 2nd memory request

The problem with this is in calculating the minimum amount of memory that the system will require. This can be difficult to estimate and many designs start with a large amount of memory, get the system running and then find out empirically the minimum amount of required memory.

In this section, the use of memory management within an embedded design has been alluded to in the case of protecting memory for corruption. While this is an important use, it is a secondary advantage compared to its ability to reuse memory through address translation. Before returning to the idea of memory protection, let's consider how address translation works and affects the memory map.

Memory management address translation

While the use of memory management usually implies the use of an operating system to remove the time-consuming job of defining and writing the driver software, it does not mean that every operating system supports memory management. Many do not or are extremely limited in the type of memory management facilities that they support. For operating systems that do support it, the designer can access standard software that controls the translation of logical addresses to different physical addresses as shown in the diagram. In this example, the processor thinks that it is accessing memory at the bottom of its memory map, while in reality it is being fetched from different locations in the main memory map. The memory does not even need to be contiguous: the processor's single block or memory can be split into smaller blocks, each with a different translation address.

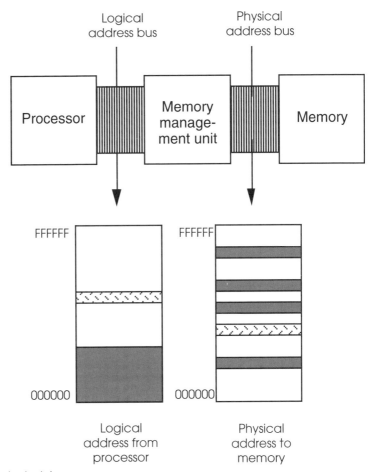

Memory management principles

This address translation is extremely powerful and allows the embedded system designer many options to provide greater fault detection and/or security within the system or even cost reduction through the use of virtual memory. The key point is that memory management divides the processor memory map into definable regions with different properties such as read and write only access for one way data transfers and task or process specific memory access.

If no memory management hardware is present, most operating systems can replace their basic address translation facility with a software-based scheme, provided code is written to be position independent and relocatable. The more sophisticated techniques start to impose a large software overhead which in many cases is hard to justify within a simple system. Address translation is often necessary to execute programs in different locations from that in which they were generated. This allows the reuse of existing software modules and facilitates the easy transfer of software from a prototype to a final system.

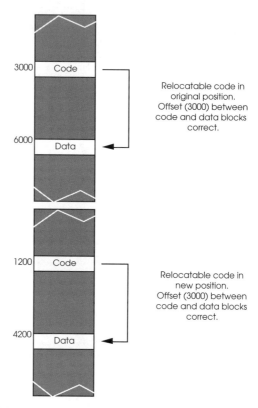

Correct movement of relocatable code

The relocation techniques are based on additional software built into the program loader or even into the operating system itself. If the operating system program loader cannot allocate the original memory, the program is relocated into the next available block and the program allowed to execute. Relocatable code does not have any immediate addressing values and makes extensive use of program relative addressing. Data areas or software subroutines are not referenced explicitly but are located by relative addressing modes using offsets:

- Explicit addressing
  e.g. branch to subroutine at address $0F04FF.

• Relative addressing

e.g. branch to subroutine which is offset from here by $50 bytes.

Provided the offsets are maintained, then the relative addressing will locate data and code wherever the blocks are located in memory. Most modern compilers will use these techniques but do not assume that all of them do.

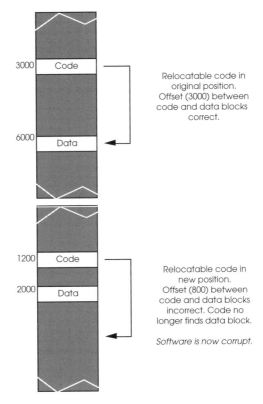

Incorrect movement of relocatable code

There are alternative ways of manipulating addresses to provide address translation, however.

Bank switching

When the first 8 bit processors became available with their 64 kbytes memory space, there were many comments concerning whether their was a need for this amount of memory. When the IBM PC appeared with its 640 kbyte memory map, the same comments were made and here we are today using PCs that probably have 32 Mbytes of RAM. The problem faced by many of the early processor architectures with a small 64 kbyte memory map is how the space can be expanded without having to change the architecture and increase register sizes and so on.

One technique that is used is that of bank switching. With this technique, additional bits are used to select different banks of memory. Each bank is the full 64 kbytes in size and is used in the normal way. By writing to the individual selection bits, an individual bank can be selected and used. It could be argued that this is no different from adding additional address bits. Using two selection bits will support four 64 kbyte banks giving a total memory space of 256 kbytes. This is the same amount of memory that can be addressed by increasing the number of address bits from 16 to 18. The difference, however, is that adding address bits implies that the programming model and processor knows about the wider address space and can use it. With bank switching, this is not the case, and the control and manipulation of the banks is under the control of the program that the processor is running. In other words, the processor itself has no knowledge that bankswitching is taking place. It simply sees the normal 64 kbyte address space.

This technique is frequently used with microcontrollers that have either small external address spaces or alternatively limited external address buses. The selection bits are created by dedicating bits from the microcontroller's parallel I/O lines and using these to select and switch the external memory banks. The bank switching is controlled by writing to these bits within the I/O port.

This has some interesting repercussions for designs that use a RTOS. The main problem is that the program must understand when the system context is safe enough to allow a bank switch to be made. This means that system entities such as data structures, buffers and anything else that is stored in memory including program software must fit into the boundaries created by the bank switching.

This can be fairly simple but it can also be extremely complex. If the bank switching is used to extend a data base, for example, the switching can be easy to control by inserting a check for a memory bank boundary. Records 1–100 could be in bank A, with bank B holding records 101–200. By checking the record number, the software can switch between the banks as needed. Such an implementation could define a subroutine as the access point to the data and within this routine the bankswitching is managed so that it is transparent to the rest of the software.

Using bank switching to support large stacks or data structures on the other hand is more difficult because the mechanisms that use the data involve both automatic and software controlled access. Interrupts can cause stacks to be incremented automatically and there is no easy way of checking for an overflow and then incorporating the bank switching and so on needed to use it.

In summary, bank switching is used and their are 8 bit processors that have dedicated bits to support it but the software issues to use it are immense. As a result, it is frequently left for the system designer to figure out a way to use it within a design. As a result, few, if any, RTOS environments support this memory model.

Segmentation

Segmentation can be described as a form of bank switching in which the processor architecture does know about! It works by providing a large external address bus but maintaining the smaller address registers and pointers within the original 8 bit architecture. To bridge the gap, the memory is segmented internally into smaller blocks that match the internal addressing and additional registers known as segment registers are used to hold the additional address data needed to complete the larger external address.

Probably the most well-known implementation is the Intel 8086 architecture.

Virtual memory

With the large linear addressing offered by today's 32 bit microprocessors, it is relatively easy to create large software applications which consume vast quantities of memory. While it may be feasible to install 64 Mbytes of RAM in a workstation, the costs are expensive compared with 64 Mbytes of a hard disk. As the memory needs go up, this differential increases. A solution is to use the disk storage as the main storage medium, divide the stored program into small blocks and keep only the blocks in processor system memory that are needed. This technique is known as virtual memory and relies on the presence within the system of a memory management unit.

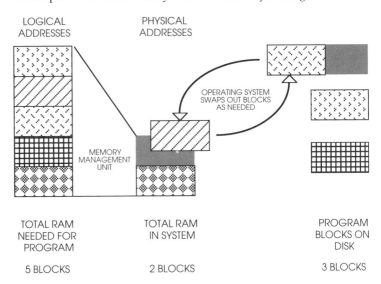

Using virtual memory to support large applications

As the program executes, the MMU can track how the program uses the blocks, and swap them to and from the disk as needed. If a block is not present in memory, this causes a page fault and forces some exception processing which performs the swapping operation. In this way, the system appears to have large amounts of system RAM when, in reality, it does not. This virtual memory technique is frequently used in workstations and in the UNIX operating system. Its appeal in embedded systems is limited because of the potential delay in accessing memory that can occur if a block is swapped out to disk.

Choosing an operating system

Comparing an operating system from 10 years ago with one offered today shows how operating system technology has developed over the past years. Although the basic functions provided by the old and the newer operating systems — they all provide multitasking, real-time responses and so on — are still present, there have been some fundamental changes in the improvement in the ease of use, performance and debugging facilities. Comparing a present-day car with one from the 1920s is a good analogy. The basic mechanics and principles have largely remained unchanged — that is, the engine, gearbox, brakes, transmission — but there has been a great improvement in the ease of driving, comfort and facilities. This is similar to what has happened with operating systems. The basic mechanisms of context switches, task control blocks, linked lists and so on are the basic fundamentals of any operating system or kernel.

As a result, it can be quite difficult to select an operating system. To make such a choice, it is necessary to understand the different ways that operating systems have developed over the years and the advantages that this has brought. The rest of this chapter discusses these changes and can help formulate the criteria that can be used to make the decision.

Assembler vs. high level language

In the early 1980s, operating systems were developed in response to minicomputer operating systems where the emphasis was on providing the facilities and performance offered by minicomputers. To achieve performance, they were often written in assembler rather than in a high level language such as C or PASCAL. The reason for this was simply one of performance: compiler technology was not advanced enough to provide the compact and fast code needed to run an operating system. For example, many compilers from the early 1980s did not use all the M68000 address and data registers and limited themselves to only one or two. The result was code that was extremely inefficient when compared with hand coded assembler which did use all the processor's registers.

The disadvantage is that assembler code is harder to write and maintain compared to a high level language and is extremely difficult to migrate to other architectures. In addition, the interface between the operating system and a high level language was not well developed and in some cases non-existent! Writing interface libraries was considered part of the software task.

As both processor performance and compiler technology improved, it became feasible to provide an operating system written in a high level language such as C which provided a seamless integration of the operating system interface and application development language.

The choice of using assembler or a high level language with some assembler compared to using an integrated operating system and high level language is fairly obvious. What was acceptable a few years ago is no longer the case and today's successful operating systems are highly integrated with their compiler technology.

ROMable code

With early operating systems, restrictions in the code development often prevented operating systems and compilers from generating code that could be blown into read only memory for an embedded application. The reasons were often historic rather than technical, although the argument that most applications were too big to fit into the relatively small size of EPROM that was available was certainly true for many years. Today, most users declare this requirement as mandatory, and it is a standard offering from compilers and operating system vendors alike.

Scheduling algorithms

One area of constant debate is that of the scheduling algorithms that are used to select which task is to execute next. There are several different approaches which can be used. The first is to switch tasks only at the end of a time slice. This allows a fairer distribution of the processing power across a large number of tasks but at the expense of response time. Another is to take the first approach but allow certain events to force switch a task even if the current one has not used up all its allotted time slice. This allows external interrupts to get a faster response. Another event that can be used to interrupt the task is an operating system call.

Others have implemented priority systems where a task's priority and status within the ready list can be changed by itself, the operating system or even by other tasks. Others have a fixed priority system where the level is fixed when the task is created. Some operating systems even allow different scheduling algorithms to be implemented so that a designer can change them to give a specific response.

Changing algorithms and so on are usually indicative of trying to squeeze the last bit of performance from the system and in such cases it may be better to use a faster processor, or even in extreme cases actually accept that the operating system cannot meet the required performance and use another.

Pre-emptive scheduling

One consistent requirement that has appeared is that of pre-emptive scheduling. This refers to a particular scheduling algorithm where the highest priority task will interrupt or pre-empt a currently executing task irrespective of whether it has used its allotted time slice, and will continue running until a higher level task is ready to. This gives the best response to interrupts and events but can be a little dangerous. If a task is given the highest priority and does not lower its priority or pre-empt itself, then other tasks will not get an opportunity to execute. Therefore the ability to pre-empt is often restricted to special tasks with time critical routines.

Modular approach

The idea of reusing code when ever possible is not a new one but it can be difficult to implement. Obvious candidates with an operating system are device drivers for I/O, and kernels for different processors. The key is in defining a standard interface which allows such modules to be reused without having to alter or change the code. This means that memory maps must not be hardwired, or assumptions made by the driver or operating system. One of the problems with early versions of many operating systems was the fact that it was not until fairly late in their development that a modular approach for device drivers was available. As a result, the standard release included several drivers for the same peripheral chip, depending on which VMEbus board it was located.

Today, this approach is no longer acceptable and operating systems are more modular in their approach and design. The advantages for users are far more compact code, shorter development times and the ability to reuse code. A special driver can be re-used without modification. This coupled with the need to keep up with the number of boards that need standard ports has led to the development of automated build systems that can take modular drivers and create a new version extremely quickly.

Re-entrant code

This follows on from the previous topic but has one fundamental difference in that a re-entrant software module can be shared be many tasks and also interrupted at any point and reused without any problems. For example, consider module A which is shared by two tasks B and C. If task B uses module A and exits from it, the module will be left in a known state, ready for another task to use it. If task C starts to use it and in the middle of its execution is switched out in favour of task B, then the problem may appear. If task B starts to use module A, it may experience problems because A will be in an indeterminate state. Even if it does not, other problems may still be lurking. If module A uses global variables, then task B will cause them to be reset. When task C returns to continue execution, its global data will have been destroyed.

A re-entrant module can tolerate such interruptions without experiencing these types of problems. The golden rule is for the module to only access data associated with the task that is using the module code. Variables are stored on stacks, in registers or in memory areas specific to the task and not to the module. If shared modules are not re-entrant, care must be taken to ensure that a context switch does not occur during its execution. This may mean disabling or locking out the scheduler or dispatcher.

Cross-development platforms

Today, most software development is done on cross-development platforms such as Sun workstations, UNIX systems and IBM PCs. This is in direct contrast to early systems which required a dedicated software development system. The degree of platform support and the availability of good development tools which go beyond the standard of symbolic level debug have become a major product selling point.

Integrated networking

This is another area which is becoming extremely important. The ability to use a network such as TCP/IP on Ethernet to control target boards, download code and obtain debugging information is fast becoming a mandatory requirement. It is rapidly replacing the more traditional method of using serial RS232 links to achieve the same end.

Multiprocessor support

This is another area which has changed dramatically. Ten years ago it was possible to use multiple processors provided the developer designed and coded all the inter-processor communication. Now, many of today's operating systems can provide optional modules that will do this automatically. However, multiprocessing techniques are often misunderstood and as this is such a big topic for both hardware and software developers it is treated in more depth in Chapters 7 and 8.

Commercial operating systems

pSOS+

pSOS+ is the name of a popular multitasking real-time operating system. Although the name refers to the kernel itself, it is often used in a more generic way to refer to a series of development tools and system components. The best way of looking at the products is to use the overall structure as shown in the diagram. The box on the left is concerned with the development environment while that on the right are the software components that are used in the final target system. The two halves work together via communication links such as serial lines, Ethernet and TCP/IP protocol or even over VMEbus itself.

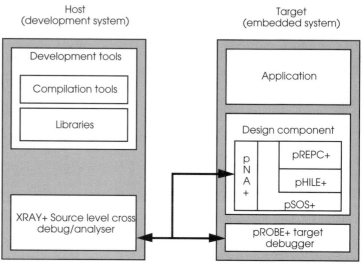

pSOS+ overall structure

pSOS+ kernel

The kernel supports a wide range of processor families like the Motorola M68000 family, the Intel 80x86 range, and the M88000 and i960 RISC processors. It is small in size and typically takes about 15–20 Kbytes of RAM, although the final figure will depend on the configuration and processor type.

It supports more than 50000 system objects such as tasks, memory partitions, message queues and so on and will execute time critical routines consistently irrespective of the application size. In other words, the time to service a message queue is the same irrespective of the size of the message. Note that this will refer to the time taken to pass the message or perform the service only and does not and cannot take into account the time taken by the user to handle messages. In other words, the consistent timing refers to the message delivery and not the actions taken as a result of the message. Worst case figures for interrupt latency and context switch for an MC68020 running at 25 MHz are 6 and 19 µs respectively.

Among its 55 service calls, it provides support for:

- Task management
- Message queues
- Event services

- Semaphore services
- Asynchronous services
- Storage allocation services
- Time management and timer services
- I/O supervisor services
- Interrupt management
- Error handling services

pSOS+m multiprocessor kernel

pSOS+m is the multiprocessing version of the kernel. From an application or task's perspective, it is virtually the same as the single processor version except that the kernel now has the ability to send and receive system objects from other processors within the system. The application or task does not know where other tasks actually reside or the communication method used to link them.

The communication mechanism works in this way. Each processor runs its own copy of the kernel and has a kernel interface to the media used for linking the processors, either a VMEbus backplane or a network. When a task sends a message to a local task, the local processor will handle it. If the message is for task running on another node, the local operating system will look up the recipient's identity and location. The message is then routed across the kernel interface across to the other processor and its operating system where the message is finally delivered.

Different kernel interfaces are needed for different media.

pREPC+ runtime support

This is a compiler independent runtime environment for C applications. It is compatible with the ANSI X3J11 technical committee's proposal for C runtime functionality and provides 88 functions that can be called from C programs. Services supported include formatted I/O, file I/O and string manipulation.

pREPC+ is not a standalone product because it uses pSOS+ or pSOS+m for device I/O and task functions and calls pHILE+ for file and disk I/O. Its routines are re-entrant which allows multiple tasks to use the same routine simultaneously.

pHILE+ file system

This product provides file I/O and will support either the MS-DOS file structure or its own proprietary formats. The MS-DOS structure is useful for data interchange while the proprietary format is designed to support the highest data throughput across a wide range of devices. pHILE+ does not drive physical devices directly but provides logical data via pSOS+ to a device driver task that converts this information to physical data and drives the storage device.

pNA+ network manager

This is a networking option that provides TCP/IP communication over a variety of media such as Ethernet and FDDI. It conforms to the UNIX 4.3 BSD socket syntax and approach and is compatible with other TCP/IP–based networking standards such as ftp and NFS.

As a result, pNA+ is used to provide efficient downloading and debugging communication between the target and a host development system. Alternatively, it can be used to provide a communication path between other systems that are also sitting on the same network.

pROBE+ system level debugger

This is the system level debugger which provides the system and low level debugging facilities. With it system objects can be inspected or even used to act as breakpoints if needed. It can use either a serial port to communicate with the outside world or if pNA+ is installed, use an TCP/IP link instead.

XRAY+ source level debugger

This is a complementary product to pROBE+ as it can use the debugger information and combine it with the C source and other symbolic information on the host to provide a complete integrated debugging environment.

OS-9

OS-9 was originally developed by Microware and Motorola as a real-time operating system for the Motorola MC6809 8 bit processor and it appeared on many 6809- based systems such as the Exorset 165 and the Dragon computer. It provided a true hierarchical filing system and the ability to run multiple tasks. It has since been ported to the M68000 family and the Intel 80x86 processor families.

OS-9 is best described as a complete operating system with its own user commands, interface and so on. Unlike other products which have concentrated on the central kernel and then built outwards but stopping at below the user and utility level, OS-9 goes from a multi-user multitasking interface with a range of utilities down to the low level kernel. Early on it supported UNIX by using and supporting the same library interface and similar system calls. So much so that one of its strengths was the ability to take UNIX source code, recompile it and then run it.

One criticism has been its poor real-time response although a new version has been released which used a smaller, compact and faster kernel to provide better performance. The full facilities are still provided by the addition of other kernel services around the inner one. It provides more sophisticated support such as multimedia extensions which other operating systems do not, and because of this and its higher level of utilities and expansion has achieved success in the marketplace.

VXWorks

VXWorks has taken another approach to the problem of providing a real-time environment as well as standard software tools and development support. Instead of creating its own or reproducing a UNIX like environment, it actually has integrated with UNIX to provide its development and operational environment. Through VXWorks' UNIX compatible networking facilities, it can combine with UNIX to form a complete runtime solution as well. The UNIX system is responsible for software development and non-real-time debugging while the VXWorks kernel is used for testing, debugging and executing the real-time applications, either standalone or part of a network with UNIX.

How does this work? The UNIX system is used as the development host and is used to edit, compile, link and administer the building of real-time code. These modules can then be burned into ROM or loaded from disk and executed standalone under the VXWorks kernel. This is possible because VXWorks can understand the UNIX object module format and has a UNIX compatible software interface. By using the standard UNIX pipe and socket mechanisms to exchange data between tasks and by using UNIX signals for asynchronous events, there is no need for recompilation or any other conversion routines. Instead, the programmer can use the same interface for both UNIX and VXWorks software without having to learn different libraries or programming commands.

Version 5.1 also offers support for POSIX 1003.4 real-time extensions and multiprocessing support for up to 20 processors is offered via another option called VxMP.

The real key to VXWorks is its ability to network with UNIX to allow a hybrid system to be developed or even allow individual modules or groups to be transferred to run in a VXWorks environment. The network can be over Ethernet or even using shared memory over VMEbus, for example.

VRTX-32

VRTX-32 from Microtec Research has gained a reputation for being an extremely compact but high performance real-time kernel. Despite being compact — typically about 8 kbytes of code for an MC68020 system — it provides task management facilities, inter-task communication, memory control and allocation, clock and timer functions, basic character I/O and interrupt handling facilities.

Under the name of VRTXvelocity, VRTX-32 systems can be designed using cross-development platforms such as Sun workstations and IBM PCs. The systems can be integrated with the host, usually using Ethernet, to provide an integrated approach to system design.

IFX

Its associated product IFX (input/output file executive) provides support for more complicated I/O subsystems such as disks, terminals, and serial communications using structures such as pipes, null devices, circular buffers and caches. The file system is MS-DOS compatible although if this is not required, disks can be treated as single partitions to speed up response.

TNX

This is the TCP/IP networking package that allows nodes to communicate with hosts and other applications over Ethernet. The Ethernet device itself can either be resident on the processor board or accessible across VMEbus. It supports both stream and datagram sockets.

RTL

This is the runtime library support for Microtec and Sun compilers and provides the library interface to allow C programs to call standard I/O functions and make VRTX-32 calls.

RTscope

This is the real-time multitasking debugger and system monitor that is used to debug VRTX tasks and applications. It operates on two levels: the board level debugger provides the standard features such as memory and register display and modify, software upload and download and so on. In the VRTX-32 system monitor mode, tasks can be interrogated, stopped, suspended and restarted.

MPV

The multiprocessor VRTX-32 extensions allow multiple processors each running their own copy of VRTX to pass messages and other task information from one processor to another and thus create a multiprocessor system. The messages are based across the VMEbus using shared memory although other links such as RS232 or Ethernet are possible.

LynxOS-POSIX conformance

POSIX (IEEE standard portable operating system interface for computer environments) began in 1986 as an attempt to provide an open standard for operating system support. The ideas behind it are to provide vendor independence, protection from technical obsolescence, the availability of standard off-the-shelf applications, the preservation of software investment and to provide connectivity between computers.

It is based on UNIX but has added a set of real-time extensions as defined in the POSIX 1003.4 document. These cover a more sophisticated semaphore system which uses the open() call to create them. This call is more normally associated with opening a file. The facilities include persistent semaphores which retain their binary state after their last use, and the ability to force a task to wait for a semaphore through a locking mechanism.

The extensions also provide a process or task locking mechanism which prevents memory pages associated with the task or process from being swapped out to memory, thus improving the real-time response for critical routines. Shared memory is better supported through the provision of the shmmap() call which will allocate a sheared memory block. Both asynchronous and synchronous I/O and inter-task message passing are supported along with real-time file extensions to speed up file I/O. This uses techniques such as preallocating file space before it is required.

At the time of writing LynxOS is the main real-time product that supports these standards, although many others support parts of the POSIX standard. Indeed, there is an increasing level of support for this type of standardisation.

However, it is not a complete panacea and, while any attempt for standardisation should be applauded, it does not solve all the issues. Real-time operating systems and applications are very different in their needs, design and approach, as can be seen from the diversity of products that are available today. Can all of these be met by a single standard? In addition, the main cost of developing software is not in porting but in testing and documenting the product and this again is not addressed by the POSIX standards. POSIX conformance means that software should be portable across processors and platforms, but it does not guarantee it. With many of today's operating systems available in versions for the major processor families, is the POSIX portability any better? Many of these questions are yet to be answered conclusively by supporters or protagonists.

An alternative way of looking at this problem is: do you assume that a ported real-time product will work because it is POSIX compliant without testing it on the new target? In most cases the answer will be no and that testing will be needed. What POSIX conformance has given is a helping hand in the right direction and this should not be belittled, neither should it be seen as a miracle cure.

In the end, the success of the POSIX standards will depend on the market and users seeing benefit in its approach. It is an approach that is gathering pace, and one that the real-time market should be aware of. It is possible that it may succeed where other attempts at a real-time interface standard have failed. Another possibility for POSIX conformance is Windows NT.

Windows NT

Windows NT has been portrayed as many different things during its short lifetime. When it first appeared, it was perceived by many as the replacement for Windows 3.1, an alternative to UNIX, and finally has settled down as an operating system for workstations, servers and power users. This chameleon like change was not due to any real changes in the product but were caused by a mixture of aspirations and misunderstandings.

Windows NT is rapidly replacing Windows 3.1 and Windows 95 and parts of its technology have already found themselves incorporated into Windows 95 and Windows for Workgroups. Whether the replacement is through a merging of the operating system technologies or through a sharing of common technology, only time will tell. The important message is that the Windows NT environment is becoming prevalent, especially with Microsoft's aim of a single set of programming interfaces that will allow an application to run on any of its operating system environments. Its greater stability and reliability is another feature that is behind its adoption by many business systems in preference over Windows 95. All this is fine, but how does this fit with an embedded system?

There are several reasons why Windows NT is being used in real-time environments. It may not have the speed of a dedicated RTOS but it has the important features and coupled with a fast processor, reasonable performance.

- Portability

    Most PC-based operating systems were written in low-level assembler language instead of a high level language such as C or C++. This decision was taken to provide smaller programs sizes and the best possible performance. The disadvantage is that the operating system and applications are now dependent on the hardware platform and it is extremely difficult to move from one platform to another. MS-DOS is written in 8086 assembler which is incompatible with the M68000 processors used in the Apple Macintosh. For a software company like Microsoft, this has an additional threat of being dependent on a single processor platform. If the platform changes — who remembers the Z80 and 6502 processors which were the mainstays of the early PCs — then its software technology becomes obsolete.

    With an operating system that is written in a high level language and is portable to other platforms, it allows Microsoft and other application developers to be less hardware dependent.

- True multitasking

    While more performant operating systems such as UNIX and VMS offer the ability to run multiple applications simultaneously, this facility is not really available from the Windows and MS-DOS environments (a full explanation of what they can do and the difference will be offered later in the chapter). This is now becoming a very important aspect for both users and developers alike so that the full performance of today's processors can be utilised.

- Multi-threaded

  Multi-threading refers to a way of creating software that can be reused without having to have multiple copies of the code or memory spaces. This leads to more efficient use of both memory and code.

- Processor independent

  Unlike Windows and MS-DOS which are completely linked to the Intel 80x86 architecture, Windows NT through its portability is processor independent and has been ported to other processor architectures such as Motorola's PowerPC, DEC's Alpha architecture and MIPS RISC processor systems.

- Multiprocessor support

  Windows NT uses a special interface to the processor hardware which makes it independent of the processor architecture that it is running on. As a result, this not only gives processor independence but also allows the operating system to run on multiprocessor systems.

- Security and POSIX support

  Windows NT offers several levels of security through its use of a multi-part access token. This token is created and verified when a user logs onto the system and contains IDs for the user, the group he is assigned to, privileges and other information. In addition, an audit trail is also provided to allow an administrator to check who has used the system, when they used it and what they did. While an overkill for a single user, this is invaluable with a system that is either used by many or connected to a network.

  The POSIX standard defines a set of interfaces that allow POSIX compliant applications to easily be ported between POSIX compliant computer systems.

  Both security and POSIX support are commercially essential to satisfy purchasing requirements from government departments, both in the US and the rest of the world.

Windows NT characteristics

Windows NT is a pre-emptive multitasking environment that will run multiple applications simultaneously and uses a priority based mechanism to determine the running order. It is capable of providing real-time support in that it has a priority mechanism and fast response times for interrupts and so on, but it is less deterministic — there is a wider range of response times — when compared to a real-time operating system such as pSOS or OS-9 used in industrial applications. It can be suitable for many real-time applications with less critical timing characteristics and this is a big advantage over the Windows 3.1 and Windows 95 environments.

Process priorities

Windows NT calls all applications, device drivers, software tasks and so on processes and this nomenclature will be used from now on. Each process can be assigned one of 32 priority levels which determines its scheduling priority.

The 32 levels are divided into two groups called the real-time and dynamic classes. The real-time classes comprise priority levels 16 through to 31 and the dynamic classes use priority levels 15 to 0. Within these two groups, certain priorities are defined as base classes and processes are allocated a base process. Independent parts of a process — these are called threads — can be assigned their own priority levels which are derived from the base class priority and can be ±2 levels different. In addition, a process cannot move from a real-time class to a dynamic one.

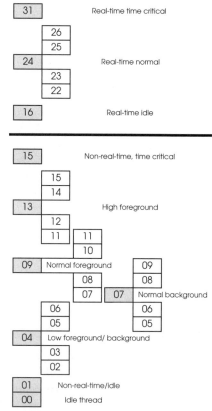

Windows NT priority levels — base class levels are shaded

The diagram shows how the base classes are organised. The first point is that within a given dynamic base class, it is possible for a lot of overlap. Although a process may have a lower base class compared to another process, it may be at a higher priority than the other one depending on the actual priority level that has been assigned to it. The real-time class is a little simpler although again there is some possibility for overlap.

User applications like word processors, spread sheets and so on run in the dynamic class and their priority will change depending on the application status. Bring an application from the background to the foreground by expanding the shrunk icon or by switching to that application will change its priority appropriately so that it gets allocated a higher priority and therefore more processing. Real-time processes include device drivers handling the keyboard, cursor, file system and other similar activities.

Interrupt priorities

The concept of priorities is not solely restricted to the preemption priority previously described. Those priorities come into play when an event or series of events occur. The events themselves are also controlled by 32 priority levels defined by the hardware abstraction layer (HAL).

Interrupt	Description
31	Hardware error interrupt
30	Powerfail interrupt
29	Inter-processor interrupt
28	Clock interrupt
27-12	Standard IBM PC AT interrupt levels 0 to 15
11-4	Reserved (not generally used)
3	Software debugger
2-0	Software interrupts for device drivers etc.

Interrupt priorities

The interrupt priorities work in a similar way to those found on a microprocessor: if an interrupt of a higher priority than the current interrupt priority mask is generated, the current processing will stop and be replaced by the associated routines for the new higher priority level. In addition, the mask will be raised to match that of the higher priority. When the higher priority processing has been completed, the previous processing will be restored and allowed to continue. The interrupt priority mask will also be restored to its previous value.

Within Windows NT, the interrupt processing is also subject to the multitasking priority levels. Depending on how these are assigned to the interrupt priority levels, the processing of a high priority interrupt may be delayed until a higher priority process has completed. It makes sense therefore to have high priority interrupts processed by processes with high priority scheduling levels. Comparing the interrupts and the priority levels shows that this maxim has been followed. Software interrupts used to communicate between processes are allocated both low interrupt and scheduling priorities. Time critical interrupts such as the clock and inter-processor interrupts are handled as real-time processes and are allocated the higher real-time scheduling priorities.

The combination of both priority schemes provides a fairly complex and flexible method of structuring how external and internal events and messages are handled.

Resource protection

If a system is going to run multiple applications simultaneously then it must be able to ensure that one application doesn't affect another. This is done through several layers of resource protection. Resource protection within MS-DOS and Windows 3.1 is a rather hit and miss affair. There is nothing to stop an application from directly accessing an I/O port or other physical device and if it did so, it could potentially interfere with another application that was already using it. Although the Windows 3.1 environment can provide some resource protection, it is of collaboration level and not mandatory. It is without doubt a case of self-regulation as opposed to obeying the rules of the system.

Protecting memory

The most important resource to protect is memory. Each process is allocated its own memory which is protected from interference by other processes through programming the memory management unit. This part of the processor's hardware tracks the executing process and ensures that any access to memory that it has not been allocated or given permission to use is stopped.

Protecting hardware

Hardware such as I/O devices are also protected by the memory management unit and any direct access is prevented. Such accesses have to be made through a device driver and in this way the device driver can control who has access to a serial port and so on. A mechanism called a spinlock is also used to control access. A process can only access a device or port if associated spinlock is not set. If it is someone else using it, the process must wait until they have finished.

Coping with crashes

If a process crashes then it is important for the operating system to maintain as much of the system as possible. This requires that the operating system as well as other applications must have its own memory and resources given to it. To ensure this is the case, processes that are specific to user applications are run in a user mode while operating system processes are executed in a special kernel mode. These modes are kept separate from each other and are protected. In addition, the operating system has to have detailed knowledge of the resources used by the crashed process so that it can clean up the process, remove it and thus free up the resources that it used. In some special cases, such as power failures where the operating system may have a limited amount of time to shut down the system in a controlled manner or save as much of the system data as it can, resources are dedicated specifically for this functionality. For example, the second highest interrupt priority is allocated to signalling a power failure.

Windows NT is very resilient to system crashes and while processes can crash, the system will continue. This is essentially due to the use of user and kernel modes coupled with extensive resource protection. Compared to Windows 3.1 and MS-DOS, this resilience is a big advantage.

Multi-threaded software

There is a third difference with Windows NT that many other operating systems do not provide in that it supports multi-threaded processes. Processes can support several independent processing paths or threads. A process may consist of several independent sections and thus form several different threads in that the context of the processing in one thread may be different from that in another thread. In other words, the process has all the resources defined that it will use and if the process can support multi-threaded operations, the scheduler will see multiple threads going through the process. A good analogy is a production line. If the production line is single threaded, it can only produce a single end product at a time. If it is multi-threaded, it separates the production process into several independent parts and each part can work on a different product. As soon as the first operation has taken place, a second thread can be started. The threads do not have to follow the same path and can vary their route through the process.

The diagram shows a simple multi-threaded operation with each thread being depicted by a different shading. As the first thread progresses through, a second thread can be started. As that progresses through, a third can commence and so on. The resources required to process the multiple threads in this case are the same as if only one thread was supported.

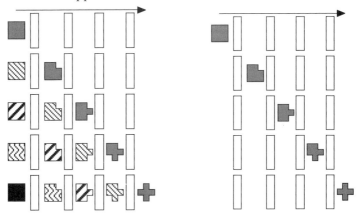

Multi-threaded (left) and single-threaded (right) operations

The advantage of multi-threaded operation is that the process does not have to be duplicated every time it is used: a new thread can be started. The disadvantage is that the process programming must ensure that there is no contention or conflict between the various threads that it supports. All the threads that exist in the process can access each other's data structures and even files. The operating system which normally polices the environment is powerless in this case. Threads within Windows NT derive their priority from that of the process although the level can be adjusted within a limited range.

Addressing space

The addressing space within Windows NT is radically different from that experienced within MS-DOS and Windows 3.1. It provides a 4 Gbyte virtual address space for each process which is linearly addressed using 32 bit address values. This is different from the segmented memory map that MS-DOS and Windows have to use. A segmented memory scheme uses 16 bit addresses to provide address spaces of only 64 kbytes. If addresses beyond this space have to be used, special support is needed to change the segment address to point to a new segment. Gone are the different types of memory such as extended and expanded.

This change towards a large 32 bit linear address space improves the environment for software applications and increases their performance and capabilities to handle large data structures. The operating system library that the applications use is called WIN32 to differentiate it from the WIN16 libraries that Windows 3.1 applications use. Applications that use the WIN32 library are known as 32 bit or even native — this term is also used for Windows NT applications that use the same instruction set as the host processor and therefore do not need to emulate a different architecture.

To provide support for legacy MS-DOS and Windows 3.1 applications, Windows NT has a 16 bit environment which simulates the segmented architecture that these applications use.

Virtual memory

The idea behind virtual memory is to provide more memory than physically present within the system. To make up the shortfall, a file or files are used to provide overflow storage for applications which are to big to fit in the system RAM at one time. Such applications' memory requirements are divided into pages and unused pages are stored on disk.

When the processor wishes to access a page which is not resident in memory, the memory management hardware asserts a page fault, selects the least used page in memory and swaps it with the wanted page stored on disk. Therefore, to reduce the system overhead, fast mass storage and large amounts of RAM are normally required.

Windows NT uses a swap file to provide a virtual memory environment. The file is dynamic in size and varies with the amount of memory that all the software including the operating system, device driver, and applications require. The Windows 3.1 swap file is limited to about 30 Mbytes in size and this effectively limits the amount of virtual memory that it can support.

The internal architecture

The internal architecture is shown in the diagram overleaf and depicts the components that run in the user and kernel modes. Most of the operating system runs in the kernel mode with the exception of the security and WIN32 subsystems.

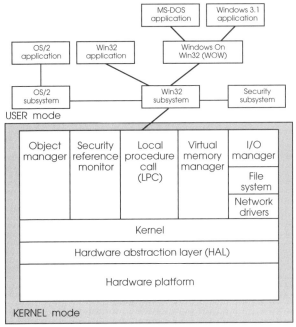

The internal Windows NT architecture

The environments are protected in that direct hardware access is not allowed and that a single application will run in a single environment. This means that some applications or combinations may not work in the Windows NT environment. On the other hand, running them in separate isolated environments does prevent them from causing problems with other applications or the operating system software.

Virtual memory manager

The virtual memory manager controls and supervises the memory requirements of an operating system. It allocates to each process a private linear address space of 4 Gbytes which is unique and cannot be accessed by other processes. This is one reason why legacy software such as Windows 3.1 applications run as if they are the only application running.

With each process running in its own address space, the virtual memory manager ensures that the data and code that the process needs to run located in pages of physical memory and ensures that the correct address translation is performed so that process addresses refer to the physical addresses where the information resides. If the physical memory is all used up, the virtual memory manager will move pages of data and code out to disk and store it in a temporary file. With some physical memory freed up, it can load from disk previously stored information and make it available for a process to use. This operation is closely associated with the scheduling process which is handled within the kernel. For an efficient operating system, it is essential to minimise the swapping out to disk — each disk swap impacts performance — and the most efficient methods involve a close correlation with process priority. Low priority processes are primary targets for moving information out to disk while high priority processes are often locked into memory so that they offer the highest performance and are never swapped out to disk. Processes can make requests to the virtual memory manager to lock pages in memory if needed.

User and kernel modes

Two modes are used to isolate the kernel and other components of the operating system from any user applications and processes that may be running. This separation dramatically improves the resilience of the operating system. Each mode is given its own addressing space and access to hardware is made through the operating system kernel mode. To allow a user process access, a device driver must be used to isolate and control its access to ensure that no conflict is caused.

The kernel mode processes use the 16 higher real-time class priority levels and thus operating system processes will take preference over user applications.

Local procedure call (LPC)

This is responsible for co-ordinating system calls from an application and the WIN32 subsystem. Depending on the type of call and to some extent its memory needs, it is possible for applications to be routed directly to the local procedure call (LPC) without going through the WIN32 subsystem.

The kernel

The kernel is responsible for ensuring the correct operation of all the processes that are running within the system. It provides the synchronisation and scheduling that the system needs. Synchronisation support takes the form of allowing threads to wait until a specific resource is available such as an object, semaphore, an expired counter or other similar entity. While the thread is waiting it is effectively dormant and other threads and processes can be scheduled to execute.

The scheduling procedures use the 32 level priority scheme previously described in this chapter and is used to schedule threads rather than processes. With a process potentially supporting multiple threads, the scheduling operates on a thread basis and not on a process basis as this gives a finer granularity and control. Not scheduling a multi-threaded process would affect several threads which may not be the required outcome. Scheduling on a thread basis gives far more control.

Interrupts and other similar events also pass through the kernel so that it can pre-empt the current thread and reschedule a higher priority thread to process the interrupt.

File system

Windows NT supports three types of file system and these different file systems can co-exist with each other although there can be some restrictions if they are accessed by non-Windows NT systems across a network, for example.

- FAT

  File allocation table is the file system used by MS-DOS and Windows 3.1 and uses file names with an 8 character name and a 3 character extension. The VFAT system used by Windows 95 and supports long file names is also supported with Windows NT v4 in that it can read Windows 95 long file names.

- HPFS

  High performance file system is an alternative file system used by OS/2 and supports file names with 254 characters with virtually none of the character restrictions that the FAT system imposes. It also uses a write caching to disk technique which stores data temporarily in RAM and writes it to disk at a later stage. This frees up an application from waiting until the physical disk write has completed. The physical disk write is performed when the processor is either not heavily loaded or when the cache is full.

- NTFS

  The NT filing system is Windows NT's own filing system which conforms to various security recommendation and allows system administrators to restrict access to files and directories within the filing system.

  All three filing systems are supported — Windows NT will even truncate and restore file names that are not MS-DOS compatible — and are selected during installation.

Network support

As previously stated, Windows NT supports most major networking protocols and through its multi-tasking capabilities can support several simultaneously using one or more network connections. The drivers that do the actual work are part of the kernel and work closely with the file system and security modules.

I/O support

I/O drivers are also part of the kernel and these provide the link between the user processes and threads and the hardware. MS-DOS and Windows 3.1 drivers are not compatible with Windows NT drivers and one major difference between Windows NT and Windows 3.1 is that not all hardware devices are supported. Typically modern devices and controllers can be used but it is wize to check the existence of a driver before buying a piece of hardware or moving from Windows 3.1 to Windows NT.

HAL approach

The hardware abstraction layer (HAL) is designed to provide portability across different processor-based platforms and between single and multi-processor systems. In essence, it defines a piece of virtual hardware that the kernel uses when it needs to access hardware or processor resources. The HAL layer then takes the virtual processor commands and requests and translates them to the actual processor system that it is actually using. This may mean a simple mapping where a Windows NT interrupt level corresponds to a processor hardware interrupt but it can involve a complete emulation of a different processor. Such is the case to support MS-DOS and Windows 3.1 applications where an Intel 80x86 processor is emulated so that an Intel instruction set can be run.

With the rest of Windows NT being written in C, a portable high level language, the only additional work to the recompilation and testing is to write a suitable HAL layer for the processor platform that is being used.

8 Writing software for embedded systems

There are several different ways of writing code for embedded systems depending on the complexity of the system and the amount of time and money that can be spent. For example, developing software for ready-built hardware is generally easier than for discrete designs. Not only are hardware problems removed — or at least they should have been — but there is more software support available to overcome the obstacles of downloading and debugging code. Many ready-built designs provide libraries and additional software support which can dramatically cut the development time.

The traditional method of writing code has centred on a two pronged approach based on the use of microprocessor emulation. The emulator would be used by the hardware designer to help debug the board before allowing the software engineer access to the prototype hardware. The software engineer would develop his code on a PC, workstation or development system, and then use the emulator as a window into the system. With the emulator, it would be possible to download code, and trace and debug it.

This approach can still be used but the ever increasing cost of emulation and the restrictions it can place on hardware design, such as timing and the physical location of the CPU and its signals, coupled with the fact that ready-built boards are proven designs, prompted the development of alternative techniques which did not need emulation. Provided a way could be found to download code and debug it on the target board, the emulator could be dispensed with. The initial solution was the addition and development of the resident onboard debugger. This has been developed into other areas and includes the development of C source and RTOS aware software simulators that can simulate both hardware and hardware on a powerful workstation. However, there is more to writing software for microprocessor-based hardware than simply compiling code and downloading it. Debugging software is covered in the next chapter.

The compilation process

When using a high level language compiler with an IBM PC or UNIX system, it is all too easy to forget all the stages that are encountered when source code is compiled into an executable file. Not only is a suitable compiler needed, but the appropriate runtime libraries and linking loader to combine all the modules are also required. The problem is that these may be well integrated for the native system, PC or workstation, but this may not be the case for a VMEbus system, where the hardware configuration may well be unique. Such cross-compilation methods, where software for another processor or target is generated on a different machine, are attractive if a suitable PC or workstation is available, but can require work to create the correct software environment. However, the popularity of this method, as opposed to the more traditional use of a dedicated development system, has increased dramatically. It is now common for operating systems to support cross-compilation directly, rather than leaving the user to piece it all together.

Compiling code

Like many compilers, such as PASCAL or C, the high level language only generates a subset of its facilities and commands from built-in routines and relies on libraries to provide the full range of functions. These libraries use the simple commands to create well-known functions, such as printf and scanf from the C language, which print and interpret data.

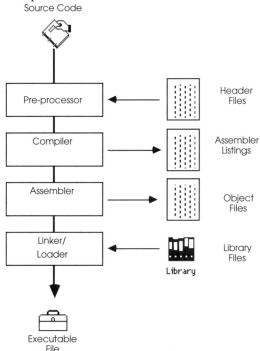

The compilation process

As a result, even a simple high level language program involves several stages and requires access to many special files.

The first stage involves pre-processing the source, where include files are added to it. These files define constants, standard functions and so on. The output of the pre-processor is fed into the compiler, where it produces an assembler file using the native instruction codes for the processor. This file may have references to other software files, called libraries. The assembler file is next assembled and converted into an object file.

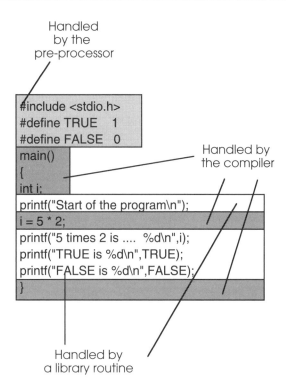

Handled
by the
pre-processor

```
#include <stdio.h>
#define TRUE    1
#define FALSE   0
main()
{
int i;
printf("Start of the program\n");
i = 5 * 2;
printf("5 times 2 is .... %d\n",i);
printf("TRUE is %d\n",TRUE);
printf("FALSE is %d\n",FALSE);
}
```

Handled by
the compiler

Handled by
a library routine

Example C source program

This contains the hexadecimal coding for the instructions, except that memory addresses and file references are not completed; these are resolved by the loader (sometimes known as a linker) that finally creates an executable file. The loader calculates all the memory addresses and takes software routines from library files to supply the standard functions called by the program.

The pre-processor

The pre-processor, as its name suggests, processes the source code before it goes through the compiler. It allows the programmer to define constants, variable types and other information. It also includes other files (*include* files) and combines them into the program source. These tasks can be conditionally performed, depending on the value of constants, and so on. The pre-processor is programmed using one of five basic commands which are inserted into the C source.

#define

#define identifier string

This statement replaces all occurrences of identifier with string. The normal convention is to put the identifier in capital letters so it can easily be recognised as a pre-processor statement.

In this example it has been used to define the values of TRUE and FALSE. The main advantage of this is usually the ability to make C code more readable. Statements like if i == 1 can be replaced by i == TRUE which makes their meaning far easier to understand. This technique is also used to define constants, which also make the code easier to understand.

One important point to remember is that the substitution is literal, i.e. the identifier is replaced by the string, irrespective of whether the substitution makes sense. While this is not usually a problem with constants, some programs use *#define* to replace part or complete program lines. If the wrong substitution or definition is made, the resulting program line may cause errors which are not immediately apparent from looking at the program lines.

It is possible to supply definitions from the C compiler command line direct to the pre-processor, without having to edit the file to change the definitions, and so on. This often allows features for debugging to be switched on or off, as required. Another use for this command is with macros.

#define MACRO() statement

#define MACRO() statement

It is possible to define a macro which is used to condense code either for space reasons or to improve its legibility. The format is *#define*, followed by the macro name and the arguments, within brackets, that it will use in the statement. There should be no space between the name and the brackets. The statement follows the bracket. It is good practice to put each argument within the statement in brackets, to ensure that no problems are encountered with strange arguments.

```
#define SQ(a) ((a)*(a))
#define MAX(i,j) ((i) > ( j) ? (i) : (j))
...
...
x = SQ(56);
z = MAX(x,y);
```

#include

#include "filename"
#include <filename>

This statement takes the contents of a file name and includes it as part of the program code. This is frequently used to define standard constants, variable types, and so on, which may be used either directly in the program source or are expected by any library routines that are used. The difference between the two forms is in the file location. If the file name is in quotation marks, the current directory is searched, followed by the standard directory — usually */usr/*

include. If angle brackets are used instead, only the standard directory is searched.

Included files are usually called header files and can themselves have further *#include* statements. The examples show what happens if a header file is not included.

#ifdef

> **#ifdef** identifier
> code
> **#else**
> code
> **#endif**

This statement conditionally includes code, depending on whether identifier has been previously defined using a *#define* statement. This is extremely useful for conditionally altering the program, depending on definitions. It is often used to insert machine dependent software into programs. In the example, the source was edited to comment out the CPU_68000 definition so that cache control information was included and a congratulations message printed. If the CPU_68040 definition had been commented out and the CPU_68000 enabled, the reverse would have happened — no cache control software is generated and an update message is printed. Note that *#ifndef* is true when the identifier does not exist and is the opposite of *#ifdef*. The *#else* and its associated code routine can be removed if not needed.

```
#define CPU_68040
/*define CPU_68000 */
#ifdef CPU_68040
/* insert code to switch on caches */
else
/* Do nothing !   */
#endif
#ifndef CPU_68040
printf("Considered upgrading to an MC68040\n");
#else
printf("Congratulations !\n");
#endif
```

#if

> **#if** expression
> code
> **#else**
> code
> **#endif**

This statement is similar to the previous *#ifdef*, except that an expression is evaluated to determine whether code is included. The expression can be any valid C expression but should be restricted to constants only. Variables cannot be used because the pre-processor does not know what values they have. This is used to assign values for memory locations and for other uses which require constants to be changed. The total memory for a program can be defined as a constant and, through a series of #*if* statements, other constants can be defined, e.g. the size of data arrays, buffers and so on. This allows the pre-processor to define resources based on a single constant and using different algorithms — without the need to edit all the constants.

Compilation

This is where the processed source code is turned into assembler modules ready for the linker to combine them with the runtime libraries. There are several ways this can be done. The first may be to generate object files directly without going through a separate assembler stage. The usual approach is to create an assembler source listing which is then run through an assembler to create an object file. During this process, it is sometimes possible to switch on automatic code optimisers which examine the code and modify it to produce higher performance.

The standard C compiler for UNIX systems is called *cc* and from its command line, C programs can be pre-processed, compiled, assembled and linked to create an executable file. Its basic options shown below have been used by most compiler writers and therefore are common to most compilers, irrespective of the platform. This procedure can be stopped at any point and options given to each stage, as needed. The options for the compiler are:

-c	Compiles as far as the linking stage and leaves the object file (suffix .o). This is used to compile programs to form part of a library.
-p	Instructs the compiler to produce code which counts the number of times each routine is called. This is the profiling option which is used with the *prof* utility to give statistics on how many subroutines are called. This information is extremely useful for finding out which parts of a program are consuming most of the processing time.
-f	Links the object program with the floating point software rather than using a hardware processor. This option is largely historic as many processors now have floating point co-processors. If the system does not, this option performs the calculations in software — but more slowly.
-g	Generates symbolic debug information for debuggers like *sdb*. Without this information, the debugger can only work at assembler level and not print variable values and so on. The symbolic information is passed through the compilation process and is stored in the executable file it produces.
-O	Switch on the code optimiser to optimise the program and improve its performance. An environment variable OPTIM controls which of two levels is used. If OPTIM=HL (high level), only the higher level code is optimised. If OPTIM=BOTH, the high level and object code optimisers are both invoked. If OPTIM is not set, only the object code optimiser is used. This option cannot be used with the -g flag.

-Wc,*args*	Passes the arguments *args* to the compiler process indicated by c, where c is one of *p012al* and stands for pre-processor, compiler first pass, compiler second pass, optimiser, assembler and linker, respectively.
-S	Compiles the named C programs and generates an assembler language output file only. This file is suffixed .s. This is used to generate source listings and allows the programmer to relate the assembler code generated by the compiler back to the original C source. The standard compiler does not insert the C source into assembler output, it only adds line references.
-E	Only runs the pre-processor on the named C programs and sends the result to the standard output.
-P	Only runs the pre-processor on the named C programs and puts the result in the corresponding files suffixed .i.
-Dsymbol	Defines a symbol to the pre-processor. This mechanism is useful in defining a constant which is then evaluated by the pre-processor, without having to edit the original source.
-Usymbol	Undefine symbol to the pre-processor. This is useful in disabling pre-processor statements.
-ldir	Provides an alternative directory for the pre-processor to find #include files. If the file name is in quotes, the pre-processor searches the current directory first, followed by dir and finally the standard directories.

Here is an example C program and the assembler listing it produced on an MC68010-based UNIX system. The assembler code uses M68000 UNIX mnemonics.

```
$cat math.c
main()
{
int a,b,c;

a=2;
b=4;
c=b-a;
b=a-c;
exit();
}
$cat math.s
file    "math.c"
   data  1
   text
   def   main; val   main; scl   2;    type  044;  endef
   global      main
main:
   ln    1
   def   ~bf;  val   ~;    scl   101;  line  2;    endef
   link.l      %fp,&F%1
#movm.l &M%1,(4,%sp)
#fmovm  &FPM%1,(FPO%1,%sp)
   def   a;    val   -4+S%1;    scl   1;    type  04;   endef
   def   b;    val   -8+S%1;    scl   1;    type  04;   endef
   def   c;    val   -12+S%1;   scl   1;    type  04;   endef
   ln    4
   mov.l &2,((S%1-4).w,%fp)
   ln    5
   mov.l &4,((S%1-8).w,%fp)
   ln    6
   mov.l ((S%1-8).w,%fp),%d1
   sub.l ((S%1-4).w,%fp),%d1
   mov.l %d1,((S%1-12).w,%fp)
```

```
ln      7
mov.l   ((S%1-4).w,%fp),%d1
sub.l   ((S%1-12).w,%fp),%d1
mov.l   %d1,((S%1-8).w,%fp)
ln      8
jsr     exit
L%12:
def     ~ef; val   ~;   scl   101;  line  9;   endef
ln      9
#fmovm  (FPO%1,%sp),&FPM%1
#movm.l (4,%sp),&M%1
unlk    %fp
rts
def     main; val  ~;   scl   -1;   endef
set     S%1,0
set     T%1,0
set     F%1,-16
set     FPO%1,4
set     FPM%1,0x0000
set     M%1,0x0000
data    1
$
```

as assembler

After the compiler and pre-processor have finished their passes and have generated an assembler source file, the assembler is used to convert this to hexadecimal. The UNIX assembler differs from many other assemblers in that it is not as powerful and does not have a large range of built-in macros and other facilities. It also frequently uses a different op code syntax from that normally used or specified by a processor manufacturer. For example, the Motorola MC68000 *MOVE* instruction becomes *mov* for the UNIX assembler. In some cases, even source and destination operand positions are swapped and some instructions are not supported. The assembler has several options:

-o objfile	Puts the assembler output into file *objfile* instead of removing the input file's .s suffix and replacing it with .o.
-n	Turns off long/short address optimisation. The default is to optimise and this causes the assembler to use short addressing modes whenever possible. The use of this option is very machine dependent.
-m	Runs the m4 macro pre-processor on the source file.
-V	Writes the assembler's version number on standard error output.

Linking and loading

On their own, object files cannot be executed as the object file generated by the assembler contains the basic program code but is not complete. The linker, or loader as it is also called, takes the object file and searches library files to find the routines it calls. It then calculates all the address references and incorporates any symbolic information. Its final task is to create a file which can be executed. This stage is often referred to as linking or loading. The linker gives the final control to the programmer concerning where sections are located in memory, which routines are used (and from which libraries) and how unresolved references are reconciled.

Symbols, references and relocation

When the compiler encounters a `printf()` or similar statement in a program, it creates an external reference which the linker interprets as a request for a routine from a library. When the linker links the program to the library file, it looks for all the external references and satisfies them by searching either default or user defined libraries. If any of these references cannot be found, an error message appears and the process aborts. This also happens with symbols where data types and variables have been used but not specified. As with references, the use of undefined symbols is not detected until the linker stage, when any unresolved or multiply defined symbols cause an error message. This situation is similar to a partially complete jigsaw, where there are pieces missing which represent the object file produced by the assembler. The linker supplies the missing pieces, fits them and makes sure that the jigsaw is complete.

The linker does not stop there. It also calculates all the addresses which the program needs to jump or branch to. Again, until the linker stage, these addresses are not calculated because the sizes of the library routines are not known and any calculations performed prior to this stage would be incorrect. What is done is to allocate enough storage space to allow the addresses to be inserted. Although the linker normally locates the program at $00000000 in memory, it can be instructed to relocate either the whole or part of the code to a different memory location. It also generates symbol tables and maps which can be used for debugging.

As can be seen, the linker stage is not only complicated but can also be extremely complex. For most compilations, the defaults used by the compiler are more than adequate.

ld linker/loader

As explained earlier, an object file generated by the assembler contains the basic program code but is not complete and cannot be executed. The command *ld* takes the object file and searches library files to find the routines it calls. It calculates all the address references and incorporates any symbolic information. Its final task is to create a COFF (common object format file) file which can be executed. This stage is often referred to as linking or loading and *ld* is often called the linker or loader. *ld* gives the final control to the programmer concerning where sections are located in memory, which routines are used (and from which libraries) and how unresolved references are reconciled. Normally, three sections are used — *.text* for the actual code, and *.data* and *.bss* for data. Again, there are several options:

-a	Produces an absolute file and gives warnings for undefined references. Relocation information is stripped from the output object file unless the option is given. This is the default if no option is specified.

-e *epsym*	Sets the start address for the output file to *epsym*.
-f *fill*	Sets the default fill pattern for holes within an output section. This is space that has not been used within blocks or between blocks of memory. The argument *fill* is a 2 byte constant.
-l*x*	Searches library libx.a, where *x* contains up to seven characters. By default, libraries are located in */lib* and */usr/lib*. The placement of this option is important because the libraries are searched in the same order as they are encountered on the command line. To ensure that an object file can extract routines from a library, the library must be searched after the file is given to the linker. Common values for *x* are *c*, which searches the standard C library and *m*, which accesses the maths library.
-m	Produces a map or listing of the input/output sections on the standard output. This is useful when debugging.
-o *outfile*	Produces an output object file called *outfile*. The name of default object file is *a.out*.
-r	Retains relocation entries in the output object file. Relocation entries must be saved if the output file is to become an input file in a subsequent *ld* session.
-s	Strips line number entries and symbol table information from the output file — normally to save space.
-t	Turns off the warning about multiply-defined symbols that are not of the same size.
-u*symname*	Enters *symname* as an undefined symbol in the symbol table.
-x	Does not preserve local symbols in the output symbol table. This option reduces the output file size.
-L*dir*	Changes the library search order so libx.a looks in *dir* before */lib* and */usr/lib*. This option needs to be in front of the -l option to work!
-N	Puts the data section immediately after the text in the output file.
-V	Outputs a message detailing the version of *ld* used.
-VS *num*	Uses *num* as a decimal version stamp to identify the output file produced.

Native vs. cross-compilers

With a native compiler, all the associated runtime libraries, default memory locations and loading software are supplied, allowing the software engineer to concentrate on writing software. It is possible to use a native compiler to write software for a target board, provided it is running the same processor. For example, it is possible to use an IBM PC compiler to write code for an embedded 80386 design or an Apple MAC compiler to create code for an M68000 target. The problem is that all the support libraries and so on must be replaced and this can be a considerable amount of work.

This is beginning to change and many compiler suppliers have realised that it is advantageous to provide many different libraries or the ability to support their development through the provision of library source. For example, the MetroWorks compilers for the MC68000 and PowerPC for the Apple MAC support cross-compilation for Windows, Windows 95 and Windows NT environments as well as embedded systems.

Run-time libraries

The first problem for any embedded design is that of runtime libraries. These provide the full range of functions that the high level language offers and can be split into several different types, depending on the functionality that they offer and the hardware that they use. The problem is that with no such thing as an embedded design, they often require some modification to get them to work.

Processor dependent

The bulk of a typical high level language library simply requires the processor and memory to execute. Mathematical functions, string manipulation, and so on, all use the processor and do not need to communicate with terminals, disk controllers and other peripherals. As a result these libraries normally require no modification. There are some exceptions concerning floating point and instruction sets. Some processors, such as the MC68020 and MC68030, can use an optional floating point co-processor while others, such as the MC68000 and MC68010, cannot. Further complications can arise between processor variants such as the MC68040 family where some have on-chip floating point, while others do not. Running floating point instructions without the hardware support can generate unexpected processor exceptions and cause the system to crash. Instruction sets can also vary, with the later generations of M68000 processors adding new codes to their predecessor's instruction set. To overcome these differences, compilers often have software switches which can be set to select the appropriate runtime to match the processor configuration.

I/O dependent

If a program does not need any I/O at all, it is very easy to move from one machine to another. However, as soon as any I/O is needed, this immediately defines the hardware that the software needs to access. Using a printf statement calls the printf routine from the appropriate library which, in turn, either drives the hardware directly or calls the operating system to perform the task of printing data to the screen. If the target hardware is different from the native target, then the printf routine will need to be rewritten to replace the native version. Any attempt to use the native version will cause a crash because either the hardware or the operating system is different.

System calls

This is a similar problem to that of I/O dependent calls. Typical routines are those which dynamically allocate memory, task control commands, use semaphores, and so on. Again, these need to be replaced with those supported by the target system.

Exit routines

These are often neglected but are essential to any conversion. With many executable files created by compilers, the program is not simply downloaded into memory and the program counter set to the start of the module. Some systems attach a module header to the file which is then used by the operating system to load the file correctly and to preload registers with address pointers to stack and heap memory and so on. Needless to say, these need to be changed or simulated to allow the file to execute on the target. The start-up routine is often not part of a library and is coded directly into the module.

Similar problems can exist with exit routines used to terminate programs. These normally use an exit() call which removes the program and frees up the memory. Again, these need to be replaced. Fortunately, the routines are normally located in the runtime library rather than being hard coded.

Writing a library

For example, given that you have an M68000 compiler running on an IBM PC and that the target is an MC68040 VMEbus system, how do you modify or replace the runtime libraries? There are two generic solutions: the first is to change the hardware design so that it looks like the hardware design that is supported by the runtime libraries. This can be quite simple and involve configuring the memory map so that memory is located at the same addresses. If I/O is used, then the peripherals must be the same — so too must their address locations to allow the software to be used without modification. The second technique is to modify the libraries so that they work with the new hardware configuration. This will involve changing the memory map, adding or changing drivers for new or different peripherals and in some cases even porting the software to a new processor or variant.

The techniques used depend on how the runtime libraries have been supplied. In some cases, they are supplied as assembler source modules and these can simply be modified. The module will have three sections: an entry and exit section to allow data to be passed to and from the routine and, sandwiched between them, the actual code that performs the requested command. It is this middle section that is modified.

Other compilers supply object libraries where the routines have already been assembled into object files. These can be very difficult to patch or modify, and in such cases the best approach is to create an alternative library.

Creating a library

The first step is to establish how the compiler passes data to routines and how it expects information to be returned. This information is normally available from the documentation or can be established by generating an assembler listing during the compilation process. In extreme cases, it may be necessary to reverse engineer the procedure using a debugger. A break point is set at the start of the routine and the code examined by hand.

The next problem is concerned with how to tell the compiler that the routine is external and needs to be specially handled. If the routine is an addition and not a replacement for a standard function, this is normally done by declaring the routines to be external when they are defined. To complement this, the routines must each have an external declaration to allow the linker to correctly match the references.

With replacements for standard library functions, the external declaration from within the program source is not needed, but the one within the replacement library routine is. The alternative library is accessed first to supply the new version by setting the library search list used by the linker.

To illustrate these procedures, consider the following PASCAL example. The first piece of source code is written in PASCAL and controls a semaphore in a typical real-time operating system, which is used to synchronise other tasks. The standard PASCAL did not have any runtime support for operating system calls and therefore a library needed to be created to supply these. The data passing mechanism is typical of most high level languages, including C and FORTRAN, and the trap mechanism, using directive numbers and parameter blocks, is also common to most operating systems.

The PASCAL program declares the operating system calls as external procedures by defining them as procedures and marking them as FORWARD. This tells the compiler and linker that they are external references that need to be resolved at the linker stage. As part of the procedure declaration, the data types that are passed to the procedure have also been defined. This is essential to force the compiler to pass the data to the routine — without it, the information will either not be accepted or the routine will misinterpret the information. In the example, four external procedures are declared: de-lay, wtsem, sgsem and atsem. The procedure delay takes an integer value while the others pass over a four character string — described as a packed array of char. Their operation is as follows:

delay   delays the task by a number of milliseconds.

atsem   creates and attaches the task to a semaphore.

wtsem   causes the task to wait for a semaphore.

sgsem   signals the semaphore.

```
program timer(input,output);

type
        datatype = packed array[1..4] of char;
```

```
var
            msecs:integer;
            name :datatype;
            i :integer;

procedure delay( msecs:integer); FORWARD;
procedure wtsem( var name:datatype); FORWARD;
procedure sgsem( var name:datatype); FORWARD;
procedure atsem( var name:datatype); FORWARD;

        begin
            name:= '1sec';
            atsem(name);
            delay(10000);
            sgsem(name);
                for i := 1 to 10 do begin;
                    wtsem(name);
                    delay(10000);
                    sgsem(name);
                end;
        end.
```

PASCAL source for the program 'TIMER'

The program TIMER works in this way. When it starts, it assigns the identity 1sec to the variable name.. This is then used to create a semaphore called 1sec using the `atsem` procedure. The task now delays itself for 10000 milliseconds to allow a second task to load itself, attach itself to the semaphore 1sec and wait for its signal. The signal comes from the `sgsem` procedure on the next line. The other task receives the signal, TIMER goes into a loop where it waits for the 1sec semaphore, delays itself for 10000 milliseconds and then signals with the 1sec semaphore. The other task complements this operation by signalling and then waiting, using the 1sec semaphore.

The end result is that the program TIMER effectively controls and synchronises the other task through the use of the semaphore.

The runtime library routines for these procedures were written in MC68000 assembler. Two of the routines have been listed to illustrate how integers and arrays are passed across the stack from a high level language — PASCAL in this case — to the routine. In C, the assembler routines would be declared as functions and the assembler modules added at link time. Again, it should be remembered that this technique is common to most compilers.

```
DELAY    IDNT  1,0
*  ++++++++++++++++++++++++++++++++++++++++++++++++++
*  ++++++++++++++++++++++++++++++++++++++++++++++++++
*  ++++ ++++
*  ++++ Runtime procedure call for PASCAL ++++
*  ++++ ++++
*  ++++ Version 1.0 ++++
*  ++++ ++++
*  ++++ Steve Heath - Motorola Aylesbury ++++
*  ++++ ++++
*  ++++++++++++++++++++++++++++++++++++++++++++++++++
```

```
*  +++++++++++++++++++++++++++++++++++++++++++++++++++
*
*          PASCAL call structure:
*
*          procedure delay(msecs:integer);FORWARD
*
*          This routine calls the delay directive of the OS
*          and delays the task for a number of ms.
*          The number is passed directly on the stack
*

           XDEF DELAY

           SECTION 9

DELAY      EQU        *
           MOVE.L     (A7)+,A4     Save return address
           MOVE.L     (A7)+,A0     Load time delay into A0
           MOVE.L     A3,-(A7)     Save A3 for PASCAL
           MOVE.L     A5,-(A7)     Save A5 for PASCAL
           MOVE.L     A6,-(A7)     Save A6 for PASCAL

EXEC       MOVE.L     #21,D0       Load directive number 21
           TRAP       #1           Execute OS command
           BNE        ERROR        Error handler if problem

POP        MOVE.L     (A7)+,A6     Restore saved values
           MOVE.L     (A7)+,A5     Restore saved values
           MOVE.L     (A7)+,A3     Restore saved values

           JMP        (A4)         Jump back to PASCAL

ERROR      MOVE.L     #14,D0       Load abort directive no.
           TRAP       #1           Abort task

           END
```

Assembler listing for the delay call

The code is divided into four parts: the first three correspond with the entry, execution and exit stages previously mentioned. A fourth part that handles any error conditions has been added.

The routine is identified to the linker as the delay procedure by the XDEF delay statement. The section 9 command instructs the linker to insert this code in the program part of the file. Note how there are no absolute addresses or address references in the source. The actual values are calculated and inserted by the linker during the linking stage.

The next few instructions transfer the data from PASCAL to the assembler routine. The return address is taken from the stack followed by the time delay. These values are stored in registers A4 and A0, respectively. Note that the stack pointer A7 is incremented after the last transfer to effectively remove the passed parameters. These are not left on the stack. The next three instructions save the address registers A3, A5 and A6 onto the stack so that they are preserved. This is necessary to successfully return to PASCAL. If they are corrupted, then the return to PASCAL will either not work or will cause the program to crash at a later point. With some compilers, more registers may need saving and it is a good idea to save all registers if it is not clear which ones must be preserved. With this example, only these three are essential.

The next part of the code loads the directive number into the right register and executes the system call using the TRAP #1 instruction. The directive needs the delay value in A0 and this is loaded earlier from the stack.

If the system call fails, the condition code register is returned with a non-zero setting. This is tested by the BNE ERROR instruction. The error routine simply executes a termination or abort system call to halt the task execution.

The final part of the code restores the three address registers and uses the return address in A4 to return to the PASCAL program. If the procedure was expecting a returned value, this would be placed on the stack using the same technique used to place the data on the stack. A common fault is to use the wrong method or fail to clear the stack of old data.

The next example routine executes the atsem directive which creates the semaphore. The assembler code is a little more complex because the name is passed across the stack using a pointer rather than the actual value and, secondly, a special parameter block has to be built to support the system call to the operating system.

```
ATSEM    IDNT  1,0
*  ++++++++++++++++++++++++++++++++++++++++++++++++++++
*  ++++++++++++++++++++++++++++++++++++++++++++++++++++
*  ++++ ++++
*  ++++ Runtime procedure call for PASCAL ++++
*  ++++ ++++
*  ++++ Version 1.0 ++++
*  ++++ ++++
*  ++++ Steve Heath - Motorola Aylesbury ++++
*  ++++ ++++
*  ++++++++++++++++++++++++++++++++++++++++++++++++++++
*  ++++++++++++++++++++++++++++++++++++++++++++++++++++
*
*          PASCAL call structure:
*
*          type
*                datatype = packed array[1..4] of char
*
*          procedure atsem(var name:datatype);FORWARD
*
*          This routine calls the OS and creates a
*          semaphore. Its name is passed across on the
*          stack using an address pointer.


           XDEF ATSEM

           SECTION 9
```

```
DELAY     EQU        *
          MOVE.L     (A7)+,A4    Save return address
          MOVE.L     (A7)+,A0    Get pointer to the name
          LEA        PBL(PC),A1  Load the PBL address
          MOVE.L     (A0),(A1)   Move the name into PBL
          MOVE.L     A3,-(A7)    Save A3 for PASCAL
          MOVE.L     A5,-(A7)    Save A5 for PASCAL
          MOVE.L     A6,-(A7)    Save A6 for PASCAL

EXEC      MOVE.L     #21,D0      Load directive number 21
          LEA        PBL(PC),A0  Load the PBL address
          TRAP       #1          Execute OS command
          BNE        ERROR       Error handler if problem

POP       MOVE.L     (A7)+,A6    Restore saved values
          MOVE.L     (A7)+,A5    Restore saved values
          MOVE.L     (A7)+,A3    Restore saved values

          JMP        (A4)        Jump back to PASCAL

ERROR     MOVE.L     #14,D0      Load abort directive no.
          TRAP       #1          Abort task

          SECTION 15

PBL       EQU        *
          DC.L       ' '    Create space for name
          DC.L       0           Semaphore key
          DC.B       0           Initial count
          DC.B       1           Semaphore type

          END
```

Assembler listing for the atsem call

The name is passed via a pointer on the stack. The pointer is fetched and then used to point to the packed array that contains the semaphore name. Normally, each byte is taken in turn by using the pointer and moving it on to the next location until it points to a null character, i.e. hexadecimal 00. Instead of writing a loop to perform this task, a short cut was taken by assuming that the name is always 4 bytes and by transferring the four characters as a single 32 bit long word.

The address of the parameter block PBL is obtained using the PC relative addressing mode. Again, the reason for this is to allow the linker freedom to locate the parameter block wherever it wants to, without the need to specify an absolute address. The address is calculated and transferred to register A1 using the load effective address instruction, LEA.

The parameter block is interesting because it has been put into section 15 as opposed to the code which is located in section 9. Both of these operations are carried out by the appropriate SECTION command. The reason for this is to ensure that the routines work in all target types, irrespective of whether there is a memory management unit present or the code is in ROM. With this compiler and linker, two sections are used for any program: section 9 is used to hold the code while section 15 is used for data. Without the section 15 command, the linker would put the parameter block immediately after the code routine somewhere in section 9. With a target with no memory management, or with it disabled, this would not cause a problem — provided the code was running in RAM. If the memory management declares the program area as read only — standard default for virtually all operating systems — or the code is in ROM, the transfer of the semaphore name would fail as the parameter block was located in read only memory. By forcing it into section 15, the block is located correctly in RAM and will work correctly, whatever the system configuration.

These routines are extremely simple and quick to create. By using a template, it is easy to modify them to create new procedure calls. More sophisticated versions could transfer all the data to build the parameter block rather than just the name, as in these examples. The procedure could even return a completion code back to the PASCAL program, if needed. In addition, register usage in these examples is not very efficient and again could be improved. However, the important point is that the amount of sophistication is dependent on what the software engineer requires.

Device drivers

This technique is not just restricted to creating runtime libraries for operating systems and replacement I/O functions. The same technique can even be used to drive peripherals or access special registers. This method creates a pseudo device driver which allows the high level language access to the lower levels of the hardware, while not going to the extreme of hard coding or in-lining assembler. If the application is moved to a different target, the pseudo device driver is changed and the application relinked with the new version.

Debugger supplied I/O routines

I/O routines which read and write data to serial ports or even sectors to and from disk can be quite time consuming to write. However, such routines already exist in the onboard debugger which is either shipped with a ready built CPU board or can be obtained for them.

```
* Output a character to console
*
* The character to be output is passed to
* this routine on the stack as byte 5 with
* reference to A7.
*
* A TRAP #14 call to the debugger does the actual work
* Tabs are handled separately

putch
 move.b 5(A7),D0 Get char from stack
 cmp #09,D0 Is it tab character?
```

```
beq _tabput Yes,go to tab routine
trap #14 Call debugger I/O
dc.w 1 Output char in D0.B
rts
```

An example putchar routine for C using debugger I/O

Many suppliers provide a list of basic I/O commands which can be accessed by the appropriate trap calls. The mechanism is very similar to that described in the previous examples: parameter block addresses are loaded into registers, the command number loaded into a data register and a trap instruction executed. The same basic technique template can be used to create replacement I/O libraries which use the debugger rather than an operating system.

Runtime libraries

The example assembler routines simply use the predefined stack mechanisms to transfer data to and from PASCAL. At no point does the routine actually know that the data is coming from a high level language as opposed to an assembler routine — let alone differentiate between C and PASCAL. If a group of high level languages has common transfer mechanisms, it should be possible to share libraries and modules between them, without having to modify them or know how they were generated. Unfortunately, this utopia has not quite been realised, although some standards have been put forward to implement it.

Using alternative libraries

Given that the new libraries have been written, how are they built into the program? This is done by the linker. The program and assembler routines are compiled and assembled into object modules. The object modules are then linked together by the linker to create the final executable program. The new libraries are incorporated using one of two techniques. The actual details vary from linker to linker and will require checking in the documentation.

Linking additional libraries

This is straightforward. The new libraries are simply included on the command line with all the other modules or added to the list of libraries to search.

Linking replacement libraries

The trick here is to use the search order so that the replacement libraries are used first instead of the standard ones. Some linkers allow you to state the search order on the command line or through a command file. Others may need several link passes, where the first pass disables the automatic search and uses the replacement library and the second pass uses the automatic search and standard library to resolve all the other calls.

Using a standard library

The reason that porting software from one environment to another is often complicated and time consuming is the difference in runtime library support. If a common set of system calls was available and only this was used by the compiler to interface to the operating system, it would be very easy to move software across from one platform to another — all that would be required would be a simple recompilation. In addition, using a common library would take advantage of common knowledge and experience.

If these improvements are so good, why is this not a more common approach? The problem is in defining the set of library calls and interface requirements. While some standards have appeared and are used, such as UNIX System V interface definition (SVID), they cannot provide a complete set for all operating system environments. Other problems can also exist with the interpretation and operation of the library calls. A call may only work with a 32 bit integer and not with an 8 or 16 bit one. Others may rely on undocumented or vaguely specified functions which may vary from one system to another, and so on. Even after taking these considerations into account, the ability to provide some standard library support is a big advantage. With it, a real-time operating system can support SVID calls and thus allow UNIX software to be transferred through recompilation with a minimum of problems.

There have been several attempts to go beyond the SVID type library definitions and provide a system library that truly supports the real-time environment. Both the VMEexec and ORKID specifications tried to implement a real-time library that was kernel independent with the plan of allowing software that used these definitions to be moved from one kernel to another. Changing kernels would allow application software to be reused with different operating system characteristics without the need to rewrite libraries and so on. The POSIX real-time definitions are another example of this type of approach.

It can be very dangerous to pin too much hope on these types of standards. The first problem is that they are source code definitions and are therefore subject to misinterpretation not only by the user, but also by the compiler, its runtime libraries and the response from the computer system itself. All of these can cause software to exhibit different behaviour. It may work on one machine but not on another. As a result, the use of a standard library does not in itself guarantee that software will work after recompilation and that it will not require major engineering effort to make it do so. What it does do, however, is provide a better base to work from and such work should be encouraged.

Porting kernels

So far, it has been assumed that the operating system or real-time kernel is already running on the target board. While this is sometimes true, it is not always the case. The operating system may not be available for the target board or the hardware may be a custom design.

Board support

One way to solve this problem is to buy the operating system software already configured for a particular board. Many software suppliers have a list of supported platforms — usually the most popular boards from the top suppliers — where their software has been ported to and is available off the shelf. For many projects, this is a very good way to proceed as it removes one more variable from the development chain. Not only do you have tested hardware, but you also have preconfigured and tested software.

Rebuilding kernels for new configurations

What happens if you cannot use a standard package or if you need to make some modifications? These changes can be made by re-building the operating system or kernel. This is not as difficult as it sounds and is more akin to a linking operation, where the various modules that comprise the operating system are linked together to form the final version.

This was not always the case. Early versions of operating systems offered these facilities but took several hours to complete and involved the study of tens of pages of tables to set various switches to include the right modules. Those of you who remember the SYSGEN command within VersaDOS will understand the problem. It did not simply link together modules, it often created them by modifying source code files and patching object files! A long and lengthy process and extremely prone to errors.

This procedure has not gone away but has become quicker and easier to manage. Through the use of reusable modules and high level languages, operating systems are modified and built using a process which is similar to compilation. User created or modified modules are compiled and then linked with the basic operating system to form the final version. The various parameters and software switches are set by various header files — similar to those used with C programs — and these control exactly what is built and where it is located.

As an example of this process, consider how VXWorks performs this task. VXWorks calls this process configuration and it uses several files to control how the kernel is configured. The process is very similar to the UNIX make command and uses the normal compilation tools used to generate tasks.

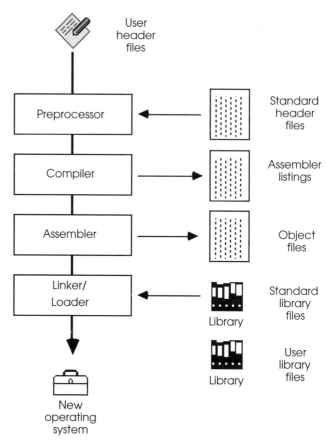

User
header
files

Standard
header
files

Preprocessor

Compiler

Assembler
listings

Assembler

Object
files

Linker/
Loader

Standard
library
files

Library

User
library
files

Library

New
operating
system

The configuration process

The three configuration files are called configAll.h, config.h and usrConfig.c. The first two files are header files, which supply parameters to the modules specified in the usrConfig.c file. Specifying these parameters without adding the appropriate statement in the usrConfig.c file will cause the build to fail.

configAll.h

This file contains all the fundamental options and parameters for kernel configurations, I/O and Networking File System parameters, optional software modules and device controllers or drivers. It also contains cache modes and addresses for I/O devices, interrupt vectors and levels.

config.h

This is where target specific parameters are stored, such as interrupt vectors for the system clock and parity errors, target specific I/O controller addresses, interrupt vectors and levels, and information on any shared memory.

usrConfig.c

This contains a series of software include statements which are used to omit or include the various software modules that the operating system may need. This file would select which Ethernet driver to use or which serial port driver was needed. These modules use parameters from the previous two configuration files within the rebuilding process.

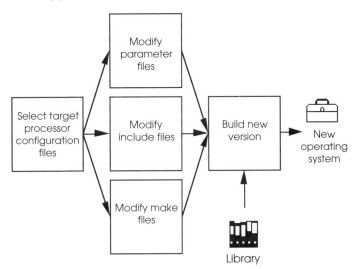

A standard build process

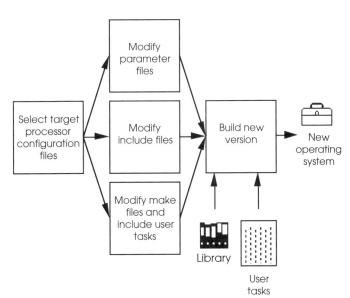

Building tasks into the operating system

Several standard make files are supplied which will create bootable, standalone versions of the operating system, as well as others that will embed tasks into the standalone version. These are used by the compiler and linker to control the building process. All the requisite files are stored in several default library directories but special directories can also be used by adding further options to the make file.

The diagrams show the basic principles involved. A standard build process usually involves modification of the normal files and simply rebuilding. New modules are extracted from the library files or directories as required to build the new version of the operating system.

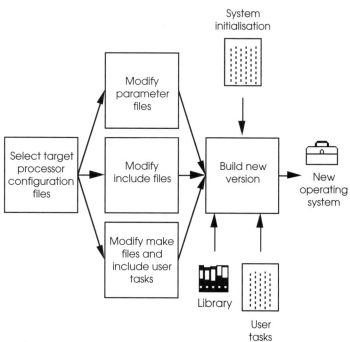

Building an embedded version

The second diagram shows the basic principles behind including user tasks into the operating system. User tasks are usually included at the make or link level and are added to the list of object files that form the operating system.

Note that this process is not the same as building an embedded standalone version. Although the tasks have been embedded, the initialisation code is still the standard one used to start up the operating system only. The tasks may be present, but the operating system is not aware of their existence. This version is often used as an intermediate stage to allow the tasks to be embedded but started under the control of a debugging shell or task.

To create a full embedded version, the user must supply some initialisation routines as well as the tasks themselves. This may involve changing the operating system start point to that of the user's own routine rather than the default operating system one. This user routine must also take care of any variable initialisation, setting up of the vector table, starting tasks in the correct order and allocating memory correctly.

Other options that may need to be included are the addition of any symbol tables for debugging purposes, multiprocessor communication and networking support.

pSOSystem+

Rebuilding operating systems is not difficult, once a basic understanding of how the process works and what needs to be changed is reached. The biggest problem faced by the user and by software suppliers is the sheer number of different parameters and drivers that are available today. With hundreds of VMEbus processor boards and I/O modules available, it is becoming extremely difficult to keep up with new product introductions. In an effort to reduce this problem, more user friendly rebuilding systems, such as pSOSystem+, are becoming available which provide a menu driven approach to the problem. The basic options are presented and the user chooses from the menu. The program then automatically generates the changes to the configuration file and builds the new version automatically.

C extensions for embedded systems

Whenever writing in a high level language, there are always times when there is a need to go down to assembler level coding, either for speed reasons or because it is simpler to do so. Accessing memory ports is another example of where this is needed. C in itself does not necessarily support this. Assembler routines can be written as library routines and included at link time — a technique that has been explained and used in this and later chapters. It is possible to define access a specific memory mapped peripheral by defining a pointer and then assigning the peripheral's memory address to the pointer. Vector tables can be created by an array of pointers to functions. While these techniques work in many cases, they are susceptible to failing.

Many compilers provide extensions to the compilers that allow embedded system software designers facilities to help them use low level code. These extensions are compiler specific and may need changing or not be supported if a different compiler is substituted. Many of these extensions are supplied as additional #pragma definitions that supply additional information to the compiler on how to handle the routines. These routines may be in C or in assembler and the number and variety will vary considerably. It is worth checking out the compiler documentation to see what it does support.

#pragma interrupt `func2`

This declares the function `func2` as an interrupt function and therefore will ensure that the compiler will save all registers so that the function code does not need to do this. It also instructs the compiler that the return mechanism is different — with a PowerPC instruction a special assembler level instruction has to be used to synchronise and restart the processor.

#pragma pure_function `func2`

This declares the function `func2` does not use or modify any global or static data and that it is a pure function. This can be used to identify assembler-based routines that configure the processor without accessing any data. This could be to change the cache control, disable or enable interrupts.

#pragma no_side_effects `func2`

This declares the function `func2` does not modify any global or static data and that it has no side effects. This could be used in preference to the pure_function option to allow access to data to allow an interrupt mask to be changed depending on a global value, for example.

#pragma no_return `func2`

This declares the function `func2` does not return and therefore the normal preparation of retaining the subroutine return address can be dispensed with. This is used when an exit or abort function is used. Jumps can also be better implemented using this as the stack will be correctly maintained and not filled with return addresses that will never be used. This can cause stack overflows.

#pragma mem_port `int2`

This declares that the variable `int2` is a value of a specific memory address and therefore should be treated accordingly. This is normally used with a definition that defines where the address is.

asm and _ _asm

The asm and _asm directives — note that the number of underlines varies from compiler to compiler — provide a way to generate assembly code from a C program. Both usually have similar functionality that allow assembler code to be directly inserted in the middle of C without having to use the external routine and linking technique. In most cases, the terms are interchangeable, but beware that this is not always the case. Care must also be taken with them as they break the main standards and enforcing strict compatibility with the compiler can cause them to either be flagged up as an error or simply ignored.

There are two ways of using the asm/_asm directives. The first is a simple way to pass a string to the assembler, an asm string. The second is an advanced method to define an asm macro that in-lines different assembly code sections, depending on the type of arguments given. The examples shown are based on the Diab PowerPC compiler.

asm strings

An asm string can be specified wherever a statement or an external declaration is allowed. It must have exactly one argument, which should be a string constant to be passed to the assembly output. Some optimisations will be turned off when an asm string statement is encountered.

```
int f() { /* returns value at $$address */
        asm(" addis r3,r0,$$address)@ha");
        asm(" lwz r3,r3,$$address@1");
```

This technique is very useful for executing small functions such as enabling and disabling interrupts, flushing caches and other processor level activities. With the code directly in-lined into the assembler, it is very quick with little or no overhead.

asm macros

An asm macro definition looks like a function definition in that the body of the function is replaced with one or more assembly code sequences. The compiler chooses one of these sequences depending on which types of arguments are provided when using the asm macro, e.g.

```
asm int busy_wait(char *addr)
        { % reg addr; lab loop;
            addi   r4,r0,1
        loop:                   # label is replaced by com-
piler
            lwarx r5,r0,addr  # argument is forced to
register
            cmpi cr0,r5,0
            bne loop
            stwcx. r4,r0,addr
            bae loop
        }

extern char *sem
fn(char *addr) {
            busy_wait(addr); /* wait for semaphore */
            busy_wait(sem) ; /* wait for semaphore */
        }
```

The first part of the source defines the assembler routine that waits for the semaphore or event to change. The second part of the source calls this assembler function twice with the event name as its parameter.

```
                    addi    r4,r0,1
        .L11:                   # label is replaced by compiler
                    lwarx       r5,r0,r31 # argument is forced to regis-
ter
                    cmpi        cr0,r5,0
                    bne         .L11
                    stwcx.      r4,r0,r31
                    bne         .L11
                    addis       r3,r0,sem@ha
                    lwz         r3,sem@1(r3)
                    addi        r4,r0,1
        .L12:                   # label is replaced by compiler
                    lwarx       5,r0,r3     # argument is forced to
register
                    cmpl        cr0,r5,0
                    bne         .L12
                    stwcx.      r4,r0,r3
                    bne         .L12
```

Downloading

Having modified libraries, linked modules together and so on, the question arises of how to get the code down to the target board. There are several methods available to do this.

Serial lines

Programs can be downloaded into a target board using a serial comms port and usually an onboard debugger. The first stage is to convert the executable program file into an ASCII format which can easily be transmitted to the target. This is done either by setting a switch on the linker to generate a download format or by using a separate utility. Several formats exist for the download format, depending on which processor family is being used. For Motorola processors, this is called the S-record format because each line commences with an S. The format is very simple and comprises of an S identifier, which describes the record type and addressing capability, followed by the number of bytes in the record and the memory address for the data. The last byte is a checksum for error checking.

```
S0080000612E6F757410
S223400600480EFFFFFFEC42AEFFF00CAE00002710FFF06C0000322D7C00000002FFFC2D27
S22340061F7C00000004FFF8222EFFF892AEFFFC2D41FFF4222EFFFC92AEFFF42D41FFF83A
S21E40063E52AEFFF06000FFC64EB9004006504E5E4E754E4B000000004E71E5
S9030000FC
```

An example S-record file

The host is then connected to the target, invokes the download command on the target debugger and sends the file. The target debugger then converts the ASCII format back into binary and loads at the correct location. Once complete, the target debugger can be used to start the program.

This method is simple but slow. If large programs need to be moved it can take all day — which is not only an efficiency problem but also leads to the practice of patching rather than updating source code. Faced with a three hour download, it is extremely tempting to go in and patch a variable or routine, rather than modify the program source, recompile and download. In practice, this method is only really suitable for small programs.

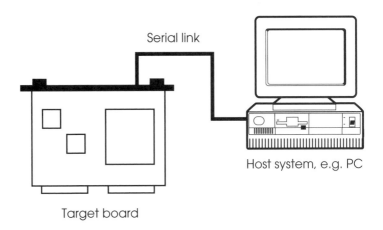

Downloading via a serial link

EPROM and FLASH

An alternative is to burn the program into EPROM, or some other form of non-volatile memory such as FLASH or battery backed-up SRAM, insert the memory chips into the target and start running the code. This can be a lot quicker than downloading via a serial line, provided the link between the development system and the PROM programmer is not a serial link itself!

There are several restrictions with this. The first is that there may not be enough free sockets on the target to accept the ROMs and second, modifications cannot be made to read only memory which means that patching and setting breakpoints will not function. If the compiler does not produce ROMable code or, for some reason, data structures have been included in the code areas, again the software may not run.

There are some solutions to this. The code in the ROMs can be block transferred to RAM before execution. This can either be done using a built-in block move command in the onboard debugger or with a small 4 or 5 line program.

Parallel ports

This is similar to the serial line technique, except that data is transferred in bytes rather than bits using a parallel interface — often a reprogrammed Centronics printer port. While a lot faster, it does require access to parallel ports which tend to be less common than serial ones.

From disk

This is an extremely quick and convenient way of downloading code. If the target is used to develop the code then this is a very easy way of downloading. If the target VMEbus board can be inserted into the development host, the code can often be downloaded directly from disk into the target memory. This technique is covered in more detail later on.

Downloading from disk can even be used with cross-compilation systems, provided the target can read floppy disks. Many target operating systems are file compatible with MS-DOS systems and use the IBM PC as their development host. In such cases, files can be transferred from the PC to the target using floppy disk(s).

Ethernet

For target systems that support networking, it is possible to download and even debug using Ethernet and TCP/IP as the communications link. This method is very common with development hosts that use UNIX. Typically, the target operating system will have a communications module which supports the TCP/IP protocols and allows it to replace a serial line for software downloading. The advantage is one of far greater transfer rates and ease of use, but it does rely on having this support available within the operating system. VXWorks, VMEexec, and pSOS+ can all cover these types of facilities.

Across a common bus

An ideal way of downloading code would be to go across a data bus such as PCI or VMEbus. This general method has already been briefly explained using an extra memory board to connect ROMs to the bus and transfer data, and the idea of adding the target boards to the host to allow the host to download directly into the target. Some operating systems can already provide this mechanism for certain host configurations. For those that do not, the methods are very simple, provided certain precautions are taken.

The first of these concerns how the operating system sees the target board. Unless restricted or told otherwise, the operating system may automatically use the memory on the target board for its own uses.

This may appear to be exactly what is required as the host can simply download code into this memory. On the other hand, the operating system may use the target memory for its own software and free up memory elsewhere in the system. Even if the memory is free, there is often no guarantee that the operating system will not overwrite the target memory.

To get around this problem, it may be necessary to physically limit the operating system so that it ignores the target memory. This can cause problems with memory management units, which will not allow access to what the operating system thinks is non-existent memory. The solution is either to disable the memory management, or to use an operating system call to get access to the physical memory to access and reserve it.

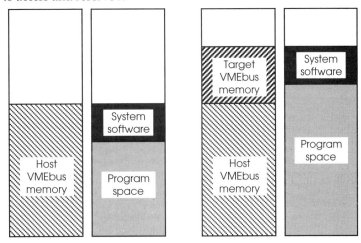

Before adding target board   After adding target board

Target memory by the operating system

With real-time-based operating systems, I would recommend disabling the MMU, thus allowing the host to access any physical memory location that the processor generates. Without the MMU, the CPU can access any address that it is instructed to, even if this address is outside those used by the operating system. This is normally not good practice but in this case the benefits justify its use. With UNIX, the best way is to declare the target memory as a shared memory segment or to access it via the device /dev/mem.

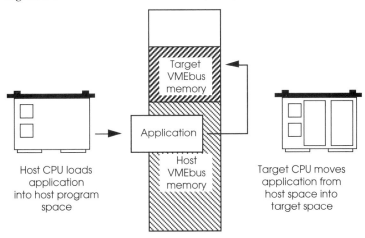

Host CPU loads application into host program space

Target CPU moves application from host space into target space

Using the target CPU to move an application to the right memory

There are occasions when even these solutions are not feasible. However, it is still possible to download software using a modification to the technique. The program is loaded into the host memory by the host. The target then moves the code across the VMEbus from the host memory space to its own memory space. The one drawback with this is the problem of memory conflicts. The target VMEbus memory must not conflict with the host VMEbus memory, and so the host must load the program into a different location to that intended. The program will have been linked to the target address, which is not recognised by the host. As a result, the host must load the program at a different physical address to that intended. This address translation function can be performed by an MMU which can translate the logical addresses of the target memory to physical addresses within the host memory space. It is usually possible to obtain the physical address by calling the operating system. This address is then used by the target to transfer the data. The alternative to this is to write a utility to simply load the program image into a specified memory location in the host memory map. After this has been done, the target CPU can transfer the program to its correct memory location.

One word of warning. It is important that there are no conflicting I/O addresses, interrupt levels, resets and so on that can cause a conflict. For example, the reset button on the target should only generate a local reset and not a VMEbus one, so that the downloading system will not see it and immediately start its reset procedure. Similarly, the processor boards should not be set up to respond to the same interrupt level or memory addresses.

This method of downloading is very quick and versatile once it has been set up. Apart from its use in downloading code during developments, the same techniques are also applicable to downloading code in multiprocessor designs.

9 Emulation and debugging techniques

Debugging techniques

The fundamental aim of a debugging methodology is to restrict the introduction of untested software or hardware to a single item. This is a good practice and of benefit to the design and implementation of any system, even those that use emulation early on in the design cycle.

It is to prevent this integration of two unknowns that simulation programs to simulate and test software, hardware or both can play a critical part in the development process.

High level language simulation

If software is written in a high level language, it is possible to test large parts of it without the need for the hardware at all. Software that does not need or use I/O or other system dependent facilities can be run and tested on other machines, such as a PC or a engineering workstation. The advantage of this is that it allows a parallel development of the hardware and software and added confidence, when the two parts are integrated, that it will work.

Using this technique, it is possible to simulate I/O using the keyboard as input or another task passing input data to the rest of the modules. Another technique is to use a data table which contains data sequences that are used to test the software.

This method is not without its restrictions. The most common mistake with this method is the use of non-standard libraries which are not supported by the target system compiler or environment. If these libraries are used as part of the code that will be transferred, as opposed to providing a user interface or debugging facility, then the modifications needed to port the code will devalue the benefit of the simulation.

The ideal is when the simulation system is using the same library interface as the target. This can be achieved by using the target system or operating system as the simulation system or using the same set of system calls. Many operating systems support or provide a UNIX compatible library which allows UNIX software to be ported using a simple recompilation. As a result, UNIX systems are often employed in this simulation role. This is an advantage which the POSIX compliant operating system Lynx offers.

This simulation allows logical testing of the software but rarely offers quantitative information unless the simulation environment is very close to that of the target, in terms of hardware and software environments.

Low level simulation

Using another system to simulate parts of the code is all well and good, but what about low level code such as initialisation routines? There are simulation tools available for these routines as well. CPU simulators can simulate a processor, memory system and, in some cases, some peripherals and allow low level assembler code and small HLL programs to be tested without the need for the actual hardware. These tools tend to fall into two categories: the first simulate the programming model and memory system and offer simple debugging tools similar to those found with an onboard debugger. These are inevitably slow, when compared to the real thing, and do not provide timing information or permit different memory configurations to be tested. However, they are very cheap and easy to use and can provide a low cost test bed for individuals within a large software team. There are even shareware simulators for the most common processors such as the one from the University of North Carolina which simulates an MC68000 processor.

```
<D0> =00000000 <D4> =00000000 <A0> =00000000 <A4> =00000000
<D1> =00000000 <D5> =0000abcd <A1> =00000000 <A5> =00000000
<D2> =00000000 <D6> =00000000 <A2> =00000000 <A6> =00000000
<D3> =00000000 <D7> =00000000 <A3> =00000000 <A7> =00000000
trace: on      sstep: on      cycles:    416     <A7'>= 00000f00
            cn tr st rc        T S  INT   XNZVC   <PC> = 00000090
   port1  00 00 82 00  SR = 1010101111011111
-------------------------------------------------------
executing a ANDI         instruction at location   58
executing a ANDI         instruction at location   5e
executing a ANDI         instruction at location   62
executing a ANDI_TO_CCR  instruction at location   68
executing a ANDI_TO_SR   instruction at location   6c
executing a OR           instruction at location   70
executing a OR           instruction at location   72
executing a OR           instruction at location   76
executing a ORI          instruction at location   78
executing a ORI          instruction at location   7e
executing a ORI          instruction at location   82
executing a ORI_TO_CCR   instruction at location   88
executing a ORI_TO_SR    instruction at location   8c
TRACE exception occurred at location   8c.
Execution halted
```

*Example display from the University of North
Carolina 68k simulator*

The second category extends the simulation to provide timing information based on the number of clock cycles. Some simulators can even provide information on cache performance, memory usage and so on, which is useful data for making hardware decisions. Different performance memory systems can be exercised using the simulator to provide performance data. This type of information is virtually impossible to obtain without using such tools. These more powerful simulators often require very powerful hosts with large amounts of memory. SDS provide a suite of such tools that can simulate a processor and memory and with some of the integrated processors that are available, even emulate onboard peripherals such as LCD controllers and parallel ports.

Simulation tools are becoming more and more important in providing early experience of and data about a system before the hardware is available. They can be a little impractical due to their performance limitations — one second of processing with a 25 MHz RISC processor taking 2 hours of simulation time was not uncommon a few years ago — but as workstation performance improves, the simulation speed increases. With instruction level simulators it is possible with a top of the range workstation to get simulation speeds of 1 to 2 MHz.

Onboard debugger

The onboard debugger provides a very low level method of debugging software. Usually supplied as a set of EPROMs which are plugged into the board or as a set of software routines that are combined with the applications code, they use a serial connection to communicate with a PC or workstation. They provide several functions: the first is to provide initialisation code for the processor and/ or the board which will normally initialise the hardware and allow it to come up into a known state. The second is to supply basic debugging facilities and, in some cases, allow simple access to the board's peripherals. Often included in these facilities is the ability to download code using a serial port or from a floppy disk.

```
>TR

PC=000404  SR=2000  SS=00A00000  US=00000000           X=0
A0=00000000  A1=000004AA  A2=00000000  A3=00000000  N=0
A4=00000000  A5=00000000  A6=00000000  A7=00A00000  Z=0
D0=00000001  D1=00000013  D2=00000000  D3=00000000  V=0
D4=00000000  D5=00000000  D6=00000000  D7=00000000  C=0
---------->LEA      $000004AA,A1

>TR

PC=00040A  SR=2000  SS=00A00000  US=00000000           X=0
A0=00000000  A1=000004AA  A2=00000000  A3=00000000  N=0
A4=00000000  A5=00000000  A6=00000000  A7=00A00000  Z=0
D0=00000001  D1=00000013  D2=00000000  D3=00000000  V=0
D4=00000000  D5=00000000  D6=00000000  D7=00000000  C=0
---------->MOVEQ    #19,D1

>
```

Example display from an onboard M68000 debugger

When the board is powered up, the processor fetches its reset vector from the table stored in EPROM and then starts to initialise the board. The vector table is normally transferred from EPROM into a

RAM area to allow it to be modified, if needed. This can be done through hardware, where the EPROM memory address is temporarily altered to be at the correct location for power-on, but is moved elsewhere after the vector table has been copied. Typically, a counter is used to determine a preset number of memory accesses, after which it is assumed that the table has been transferred by the debugger and the EPROM address can safely be changed.

The second method, which relies on processor support, allows the vector table to be moved elsewhere in the memory map. With the later M68000 processors, this can also be done by changing the vector base register which is part of the supervisor programming model.

The debugger usually operates at a very low level and allows basic memory and processor register display and change, setting RAM-based breakpoints and so on. This is normally performed using hexadecimal notation, although some debuggers can provide a simple disassembler function. To get the best out of these systems, it is important that a symbol table is generated when compiling or linking software, which will provide a cross-reference between labels and symbol names and their physical address in memory. In addition, an assembler source listing which shows the assembler code generated for each line of C or other high level language code is invaluable. Without this information it can be very difficult to use the debugger easily. Having said that, it is quite frustrating having to look up references in very large tables and this highlights one of the restrictions with this type of debugger.

While considered very low level and somewhat limited in their use, onboard debuggers are extremely useful in giving confidence that the board is working correctly and working on an embedded system where an emulator may be impractical. However, this ability to access only at a low level can also place severe limitations on what can be debugged.

The first problem concerns the initialisation routines and in particular the processor's vector table. Breakpoints use either a special breakpoint instruction or an illegal instruction to generate a processor exception when the instruction is executed. Program control is then transferred to the debugger which displays the breakpoint and associated information. Similarly, the debugger may use other vectors to drive the serial port that is connected to the terminal.

This vector table may be overwritten by the initialisation routines of the operating system which can replace them with its own set of vectors. The breakpoint can still be set but when it is reached, the operating system will see it instead of the debugger and not pass control back to it. The system will normally crash because it is not expecting to see a breakpoint or an illegal instruction!

To get around this problem, the operating system may need to be either patched so that its initialisation routine writes the debugger vector into the appropriate location or this must be done using the debugger itself. The operating system is single stepped through its initialisation routine and the instruction that overwrites the vector simply skipped over, thus preserving the debugger's vector. Some operating systems can be configured to preserve the debugger's exception vectors, which removes the need to use the debugger to preserve them.

A second issue is that of memory management where there can be a problem with the address translation. Breakpoints will still work but the addresses returned by the debugger will be physical, while those generated by the symbol table will normally be logical. As a result, it can be very difficult to reconcile the physical address information with the logical information.

The onboard debugger provides a simple but sometimes essential way of debugging VMEbus software. For small amounts of code, it is quite capable of providing a method of debugging which is effective, albeit not as efficient as a full blown symbolic level debugger — or as complex or expensive. It is often the only way of finding out about a system which has hung or crashed.

Task level debugging

In many cases, the use of a low level debugger is not very efficient compared with the type of control that may be needed. A low level debugger is fine for setting a breakpoint at the start of a routine but it cannot set them for particular task functions and operations. It is possible to set a breakpoint at the start of the routine that sends a message, but if only a particular message is required, the low level approach will need manual inspection of all messages to isolate the one that is needed — an often daunting and impractical approach!

To solve this problem, most operating systems provide a task level debugger which works at the operating system level. Breakpoints can be set on system circumstances, such as events, messages, interrupt routines and so on, as well as the more normal memory address. In addition, the ability to filter messages and events is often included. Data on the current executing tasks is provided, such as memory usage, current status and a snapshot of the registers.

Symbolic debug

The ability to use high level language instructions, functions and variables instead of the more normal addresses and their contents is known as symbolic debugging. Instead of using an assembler listing to determine the address of the first instruction of a C function and using this to set a breakpoint, the symbolic debugger allows the breakpoint to be set by quoting a line reference or the function name. This interaction is far more efficient than working at the assembler level, although it does not necessarily mean losing the ability to go down to this level if needed.

The reason for this is often due to the way that symbolic debuggers work. In simple terms, they are intelligent front ends for assembler level debuggers, where software performs the automatic look-up and conversion between high level language structures and their respective assembler level addresses and contents.

```
12 int prime,count,iter;
13
14 for (iter = 1;iter<=MAX_ITER;iter++)
15 {
16 count = 0;
17 for(i = 0; i<MAX_PRIME; i++)
18 flags[i] = 1;
19 for(i = 0; i<MAX_PRIME; i++)
20 if(flags[i])
21 {
22 prime = i + i + 3;
23 k = i + prime;
24 while (k < MAX_PRIME)
25 {
26 flags[k] = 0;
27 k += prime;
28 }
29 count++;
```

Source code listing with line references

```
000100AA 7C01 MOVEQ #$1,D6
000100AC 7800 MOVEQ #$0,D4
000100AE 7400 MOVEQ #$0,D2
000100B0 207C 0001 2148 MOVEA.L #$12148,A0 000100B6 11BC
0001 2000 MOVE.B #$1,($0,A0,D2.W)
000100BC 5282 ADDQ.L #$1,D2
000100BE 7011 MOVEQ #$11,D0
000100C0 B082 CMP.L D2,D0
000100C2 6EEC BGT.B $100B0
000100C4 7400 MOVEQ #$0,D2 000100C6 207C 0001 2148 MOVEA.L
#$12148,A0 000100CC 4A30 2000 TST.B ($0,A0,D2.W) 000100D0
6732 BEQ.B $10104 000100D2 2A02 MOVE.L D2,D5 000100D4 DA82
ADD.L D2,D5 000100D6 5685 ADDQ.L #$3,D5
```

Assembler listing

```
>>> 12 int prime,count,iter;
>>> 13
>- 14 => for (iter = 1;<=iter<=MAX_ITER;iter++)
> 000100AA 7C01 MOVEQ #$1,D6
>>> 15 {
>>> 16 count = 0;
> 000100AC 7800 MOVEQ #$0,D4
>- 17 => for(i = 0;<= i<MAX_PRIME; i++)> > 000100AE 7400
MOVEQ #$0,D2
>>> 18 flags[i] = 1;
> 000100B0 207C 0001 2148 MOVEA.L #$12148,A0 {flags}
> 000100B6 11BC 0001 2000 MOVE.B #$1,($0,A0,D2.W)
>- 17 for(i = 0; i<MAX_PRIME; => i++)<=
> 000100BC 5282 ADDQ.L #$1,D2
>- 17 for(i = 0; => i<MAX_PRIME;<=i++)
> 000100BE 7011 MOVEQ #$11,D0
> 000100C0 B082 CMP.L D2,D0
> 000100C2 6EEC BGT.B $100B0
```

Assembler listing with symbolic information

The key to this is the creation of a symbol table which provides the cross-referencing information that is needed. This can either be included within the binary file format used for object and absolute files or, in some cases, stored as a separate file. The important thing to remember is that symbol tables are often not automatically created and, without them, symbolic debug is not possible.

When the file or files are loaded or activated by the debugger, it searches for the symbolic information which is used to display more meaningful information as shown in the various listings. The symbolic information means that breakpoints can be set on language statements as well as individual addresses. Similarly, the code can be traced or stepped through line by line or instruction by instruction.

This has several repercussions. The first is the number of symbolic terms and the storage they require. Large tables can dramatically increase file size and this can pose constraints on linker operation when building an application or a new version of an operating system. If the linker has insufficient space to store the symbol tables while they are being corrected — they are often held in RAM for faster searching and update — the linker may crash with a symbol table overflow error. The solution is to strip out the symbol tables from some of the modules by recompiling them with symbolic debugging disabled or by allocating more storage space to the linker.

The problems may not stop there. If the module is then embedded into a target and symbolic debugging is required, the appropriate symbol tables must be included in the build and this takes up memory space. It is not uncommon for the symbol tables to take up more space than the spare system memory and prevent the system or task from being built or running correctly. The solution is to add more memory or strip out the symbol tables from some of the modules.

It is normal practice to remove all the symbol table information from the final build to save space. If this is done, it will also remove the ability to debug using the symbol information. It is a good idea to have at least a hard copy of the symbol table to help should any debugging be needed.

Emulation

Even using the described techniques, it cannot be stated that there will never be a need for additional help. There will be times when instrumentation, such as emulation and logic analysis, is necessary to resolve problems within a design quickly. Timing and intermittent problems cannot be easily solved without access to further information about the processor and other system signals. Even so, the recognition of a potential problem source, such as a specific software module or hardware, allows more productive use and a speedier resolution. The adoption of a methodical design approach and the use of ready built boards as the final system, at best remove the need for emulation and, at worst, reduce the amount of time debugging the system.

There are some problems with using emulation within a board-based system or any rack mounted system. The first is how to get the emulation or logic analysis probe onto the board in the first place. Often the gap between the processor and adjacent boards is too small to cope with the height of the probe. It may be possible to move adjacent boards to other slots, but this can be very difficult or impossible in densely populated racks. The answer is to use an extender board to move the target board out of the rack for easier access. Another problem is the lack of a socketed processor chip which effectively prevents the CPU from being removed and the emulator probe from being plugged in. With the move towards surface mount and high pin count packages, this problem is likely to increase. If you are designing your own board, I would recommend that sockets are used for the processor to allow an emulator to be used. If possible, and the board space allows it, use a zero insertion force socket. Even with low insertion force sockets, the high pin count can make the insertion force quite large. One option that can be used, but only if the hardware has been designed to do so, is to leave the existing processor *in situ* and tri-state all its external signals. The emulator is then connected to the processor bus via another connector or socket and takes over the processor board.

The second problem is the effect that large probes can have on the design especially where high speed buses are used. Large probes and the associated cabling create a lot of additional capacitance loading which can prevent an otherwise sound electronic design from working. As a result, the system speed very often must be downgraded to compensate. This means that the emulator can only work with a slower than originally specified design. If there is a timing problem that only appears while the system is running at high speed, then the emulator is next to useless in providing any help. We will come back to emulation techniques at the end of this chapter.

Optimisation problems

The difficulties do not stop with hardware mechanical problems. Software debugging can be confused or hampered by optimisation techniques used by the compiler to improve the efficiency of the code. Usually set by options from the command line, the optimisation routines examine the code and change it to improve its efficiency, while retaining its logical design and context. Many different techniques are used but they fall into two main types: those that remove code and those that add code or change it. A compiler may remove variables or routines that are never used or do not return any function. Small loops may be unrolled into straight line code to remove branching delays at the expense of a slightly larger program. Floating point routines may be replaced by inline floating point instructions. The net result is code that is different from the assembler listing produced by the compiler. In addition, the generated symbol table may be radically different from that expected from the source code.

These optimisation techniques can be ruthless: I have known whole routines to be removed and in one case a complete program was reduced to a single NOP instruction! The program was a set of functions that performed benchmark routines but did not use any global information or return any values. The optimiser saw this and decided that as no data was passed to it and it did not modify or return any global data, it effectively did nothing and replaced it with a NOP. When benchmarked, it gave a pretty impressive performance of zero seconds to execute several million calculations.

```c
/* sieve.c – Eratosthenes Sieve prime number calculation
*/
/* scaled down with MAX_PRIME set to 17 instead of 8091 */


#define MAX_ITER    1
#define MAX_PRIME   17

c
har      flags[MAX_PRIME];

main ()
{
        register int i,k,l,m;
        int    prime,count,iter;

        for (iter = 1;iter<=MAX_ITER;iter++)
            {
            count = 0;
/* redundant code added here */
            for(l = 0; l < 200; l++ );
            for(m = 128; l > 1; m— );
/* redundant code ends here */
            for(i = 0; i<MAX_PRIME; i++)
                flags[i] = 1;
            for(i = 0; i<MAX_PRIME; i++)
                if(flags[i])
                    {
                    prime = i + i + 3;
                    k = i + prime;
                    while (k < MAX_PRIME)
                        {
                        flags[k] = 0;
                        k += prime;
                        }
                    count++;
                    printf(" prime %d = %d\n",
count, prime);
                    }
            }
        printf("\n%d primes\n",count);
}
```

Source listing for optimisation example

```
                   file      "ctm1AAAa00360"                         file      "ctm1AAAa00355"

                   def       aut1.,32                                def       aut1.,32
                   def       arg1.,64                                def       arg1.,56
                   text                                              text
                   global    _main                                  global    _main
_main:                                                   _main:
                   subu      r31,r31,arg1.                           subu      r31,r31,arg1.
                   st        r1,r31,arg1.-4                          st        r1,r31,arg1.-4
                   st        r19,r31,aut1.+0                         st.d      r20,r31,aut1.+0
                   st        r20,r31,aut1.+4                         st.d      r22,r31,aut1.+8
                   st        r21,r31,aut1.+8                         st        r25,r31,aut1.+16
                   st        r22,r31,aut1.+12                        or        r20,r0,1
                   st        r23,r31,aut1.+16            @L26:
                   st        r24,r31,aut1.+20
                   st        r25,r31,aut1.+24                        or        r21,r0,r0
                   or        r19,r0,1                                or        r25,r0,r0
                   br        @L25                        @L7:
@L26:
                                                                     addu      r25,r25,1
                   or        r20,r0,r0                               cmp       r13,r25,200
                   or        r23,r0,r0                               bbl       lt,r13,@L7
                   br        @L6                                     br.n      @L28
@L7:                                                                 or        r2,r0,128
                   addu      r23,r23,1                   @L11:
@L6:                                                                 subu      r2,r2,1
                   cmp       r13,r23,200                 @L28:
                   bbl       lt,r13,@L7                              cmp       r13,r25,1
                   or        r22,r0,128                              bbl       gt,r13,@L11
                   br        @L10                                    or        r25,r0,r0
@L11:                                                                or.u      r22,r0,hi16(_flags)
                   subu      r22,r22,1                               or        r22,r22,lo16(_flags)
@L10:
                   cmp       r13,r23,1                   @L15:
                   bbl       gt,r13,@L11                             or        r13,r0,1
                   or        r25,r0,r0                               st.b      r13,r22,r25
                   br        @L14                                    addu      r25,r25,1
@L15:                                                                cmp       r12,r25,17
                   or.u      r13,r0,hi16(_flags)                     bbl       lt,r12,@L15
                   or        r13,r13,lo16(_flags)                    or        r25,r0,r0
                   or        r12,r0,1                    @L24:
                   st.b      r13,r22,r25                             ld.b      r12,r22,r25
                   addu      r25,r25,1                               bcnd      eq0,r12,@L17
@L14:                                                                addu      r12,r25,r25
                   cmp       r13,r25,17                              addu      r23,r12,3
                   bbl       lt,r13,@L15                             addu      r2,r25,r23
                   or        r25,r0,r0                               cmp       r12,r2,17
                   br        @L23                                    bbl       ge,r12,@L18
@L24:
                   or.u      r13,r0,hi16(_flags)         @L20:
                   or        r13,r13,lo16(_flags)                    st.b      r0,r22,r2
                   ld.b      r13,r13,r25                             addu      r2,r2,r23
                   bcnd      eq0,r13,@L17                            cmp       r13,r2,17
                   addu      r13,r25,r25                             bbl       lt,r13,@L20
                   addu      r21,r13,3                   @L18:
                   addu      r24,r25,r21
                   br        @L19                                    addu      r21,r21,1
@L20:                                                                or.u      r2,r0,hi16(@L21)
                   or.u      r13,r0,hi16(_flags)                     or        r2,r2,lo16(@L21)
                   or        r13,r13,lo16(_flags)                    or        r3,r0,r21
                   st.b      r0,r13,r24                              bsr.n     _printf
                   addu      r24,r24,r21                             or        r4,r0,r23
@L19:                                                    @L17:
                   cmp       r13,r24,17                              addu      r25,r25,1
                   bbl       lt,r13,@L20                             cmp       r13,r25,17
                   addu      r20,r20,1                               bbl       lt,r13,@L24
                   or.u      r2,r0,hi16(@L21)                        addu      r20,r20,1
                   or        r2,r2,lo16(@L21)                        cmp       r13,r20,1
                   or        r3,r0,r20                               bbl       le,r13,@L26
                   or        r4,r0,r21                               or.u      r2,r0,hi16(@L27)
                   bsr       _printf                                 or        r2,r2,lo16(@L27)
@L17:                                                                bsr.n     _printf
                   addu      r25,r25,1                               or        r3,r0,r21
@L23:                                                                ld.d      r20,r31,aut1.+0
                   cmp       r13,r25,17                              ld        r1,r31,arg1.-4
                   bbl       lt,r13,@L24                             ld.d      r22,r31,aut1.+8
                   addu      r19,r19,1                               ld        r25,r31,aut1.+16
@L25:                                                                jmp.n     r1
                   cmp       r13,r19,1                               addu      r31,r31,arg1.
                   bbl       le,r13,@L26
                   or.u      r2,r0,hi16(@L27)
                   or        r2,r2,lo16(@L27)
                   or        r3,r0,r20
                   bsr       _printf
                   ld        r19,r31,aut1.+0
                   ld        r20,r31,aut1.+4
                   ld        r21,r31,aut1.+8
                   ld        r22,r31,aut1.+12
                   ld        r23,r31,aut1.+16
                   ld        r24,r31,aut1.+20
                   ld        r25,r31,aut1.+24
                   ld        r1,r31,arg1.-4
                   addu      r31,r31,arg1.
                   jmp       r1
```

No optimisation   Full optimisation

Assembler listings for optimised and non-optimised compilation

To highlight how optimisation can dramatically change the
generated code structure, look at the C source listing for the
Eratosthenes Sieve program and the resulting M88000 assembler

listings that were generated by using the default non-optimised setting and the full optimisation option. The immediate difference is in the greatly reduced size of the code and the use of the .n suffix with jump and branch instructions to make use of the delay slot. This is a technique used on many RISC processors to prevent a pipeline stall when changing the program flow. If the instruction has a .n suffix, the instruction immediately after it is effectively executed with the branch and not after it, as it might appear from the listing!

In addition, the looping structures have been reorganised to make them more efficient, although the redundant code loops could be encoded simply as a loop with a single branch. If the optimiser is that good, why has it not done this? The reason is that the compiler expects loops to be inserted for a reason and usually some form of work is done within the loop which may change the loop variables. Thus the compiler will take the general case and use that rather than completely remove it or rewrite it. If the loop had been present in a dead code area — within a conditional statement where the conditions would never be met — the compiler would remove the structure completely.

The initialisation routine _main is different in that not all the variables are initialised using a store instruction and fetching their values from a stack. The optimised version uses the faster 'or' instruction to set some of the variables to zero.

These and other changes highlight several problems with optimisation. The obvious one is with debugging the code. With the changes to the code, the assembler listing and symbol tables do not match. Where the symbols have been preserved, the code may have dramatically changed. Where the routines have been removed, the symbols and references may not be present. There are several solutions to this. The first is to debug the code with optimisation switched off. This preserves the symbol references but the code will not run at the same speed as the optimised version, and this can lead to some timing problems. A second solution is becoming available from compiler and debugger suppliers, where the optimisation techniques preserve as much of the symbolic information as possible so that function addresses and so on are not lost.

The second issue is concerned with the effect optimisation may have on memory mapped I/O. Unless the optimiser can recognise that a function is dealing with memory mapped I/O, it may not realise that the function is doing some work after all and remove it — with disastrous results. This may require declaring the I/O addresses as a global variable, returning a value at the function's completion or even passing the address to the function itself, so that the optimiser can recognise its true role. A third complication can arise with optimisations such as unrolling loops and software timing. It is not uncommon to use instruction sequences to delay certain accesses or functions. A peripheral may require a certain number of clock cycles to respond to a command. This delay can be accomplished by executing other instructions, such as a loop or a divide instruction. The optimiser may remove or unroll such loops and replace the inefficient divide instruction with a logical shift. While this does increase the performance, that is not what was required and the delayed peripheral access may not be long enough — with disastrous results.

Such software timing should be discouraged not only for this but also for portability reasons. The timing will assume certain characteristics about the processor in terms of processing speed and performance which may not be consistent with other faster board designs or different processor versions.

Xray

It is not uncommon to use all the debugging techniques that have been described so far at various stages of a development. While this itself is not a problem, it has been difficult to get a common set of tools that would allow the various techniques to be used without having to change compilers or libraries, learn different command sets, and so on. The ideal would be a single set of compiler and debugger tools that would work with a simulator, task level debugger, onboard debugger and emulator. This is exactly the idea behind Microtec's Xray product.

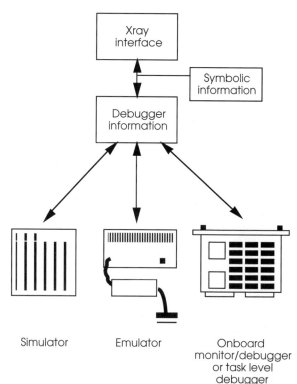

Xray structure

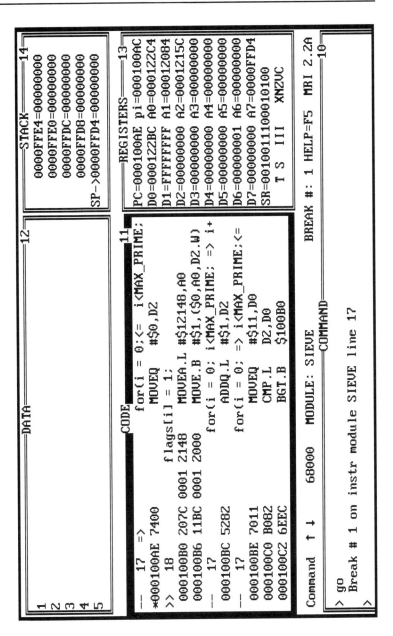

Xray screen shot

Xray consists of a consistent debugger system that can interface with a simulator, emulator, onboard debugger or operating system task level debugger. It provides a consistent interface which greatly improves the overall productivity because there is no relearning required when moving from one environment to another. It obtains its debugging information from a variety of sources, depending on how the target is being accessed. With the simulator, the information is accessed directly. With an emulator or target hardware, the link is via a simple serial line, via Ethernet or directly across a shared memory interface. The data is then used in conjunction with symbolic information to produce the data that the user can see and control on the host machine.

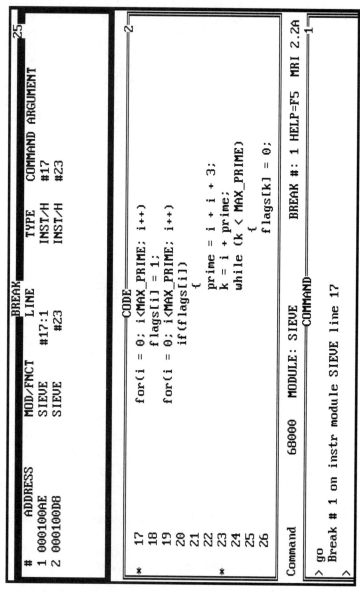

An Xray screen shot

The interface consists of a number of windows which display the debugging information. The windows consist of two types: those that provide core information, such as breakpoints and the processor registers and status. The second type are windows concerned with the different environments, such as task level information. Windows can be chosen and displayed at the touch of a key.

The displays are also consistent over a range of hosts, such as Sun workstations, IBM PCs and UNIX platforms. Either a serial or network link is used to transfer information from the target to the debugger. The one exception is that of the simulator which runs totally on the host system.

So how are these tools used? Xray comes with a set of compiler tools which allows software to be developed on a host system. This system does not have to use the same processor as the target. To execute the code, there is a variety of choices. The simulator is ideal for debugging code at a very early stage, before hardware is available, and allows software development to proceed in parallel with hardware development. Once the hardware is available, the Xray interface can be taken into the target through the use of an emulator or a small onboard debug monitor program. These debug monitors are supplied as part of the Xray package for a particular processor. They can be easily modified to reflect individual memory maps and have drivers from a large range of serial communications peripherals.

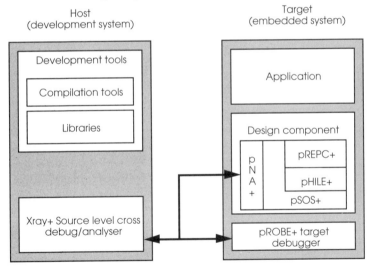

Host
(development system)

Target
(embedded system)

pSOS⁺ debugging

With Xray running in the target, the hardware and initial software routines can be debugged. The power of Xray can be further extended by having an Xray interface from the operating system debugger. pSOS⁺ uses this method to provide its debugging interface. This allows task level information to be used to set breakpoints, and so on, while still preserving the lower level facilities. This provides an extremely powerful and flexible way of debugging a target system. Xray has become a *de facto* standard for debugging tools within the real-time and VMEbus market. This type of approach is also being adopted by many other software suppliers.

MasterWorks

Microtec's MasterWorks product takes the idea of Xray further. It uses the basic technique of having a window in the target system through which the debug information can be extracted and transferred to an intelligent user interface, but extends the information that it obtains. It can profile software to see how much time is spent in various functions and it can build flow diagrams based on program execution as well as offer the existing Xray facilities. This gives the software engineer very sophisticated tools to optimise and improve as well as debug the system and application software.

The role of the development system

An alternative environment for developing software for embedded systems is to use the final operating system as the development environment either on the target system itself or on a similar system. This used to be the traditional way of developing software before the advent of cheap PCs and workstations and integrated cross-compilation.

It still offers distinct advantages over the cross-compilation system. Runtime library support is integrated because the compilers are producing native code. Therefore the runtime libraries that produce executable code on the development system will run unmodified on the target system. The final software can even be tested on the development system before moving it to the target. In addition, the full range of functions and tools can be used to debug the software during this testing phase, which may not be available on the final target. For example, if a target simply used the operating system kernel, it would not have the file system and terminal drivers needed to support an onscreen debugger or help download code quickly from disk to memory. Yet a development system running the full version of the operating system could offer this and other features, such as downloading over a network.

However, there are some important points to remember.

Floating point and memory management functions

Floating point co-processors and memory management units should be considered as part of the processor environment and need to be considered when creating code on the target. For example, the development system may have a floating point unit and memory management, while the target does not. Code created on the development system may not run on the target because of these differences. Executing floating point instructions would cause a processor exception while the location of code and data in memory may not be compatible.

This means that code created for the development system may need recompiling and linking to ensure that the correct runtime routines are used for the target and that they are located correctly.

This in turn may mean that the target versions may not run on the development system because its resources do not match up. This raises the question of the validity of using a development system in the first place. The answer is that the source code and the bulk of the binary code does not need modifying. Calling up a floating point emulation library instead of using floating point instructions will not affect any integer or I/O routines. Linking modules to a different address changes the addresses, not the instructions and so the two versions are still extremely similar. If the code works on the development system, it is likely that it will work on the target system.

While the cross-compilation system is probably the most popular method used today to develop software for embedded systems — due to the widespread availability of PCs and workstations and the improving quality of software tools — it is not the only way. Dedicated development systems can offer faster and easier software creation because of the closer relationship between the development environment and the end target.

Emulation techniques

In-circuit emulation (ICE) has been the traditional method employed to emulate a processor inside an embedded design so that software can be downloaded and debugged *in situ* in the end application. For many processors this is still an appropriate method for debugging embedded systems but the later processors have started to dispense with the emulator as a tool and replace it with alternative approaches.

The main problem is concerned with the physical issues associated with replacing the processor with a probe and cable. These issues have been touched on before but it is worth revisiting them. The problems are:

- Physical limitation of the probe

  With high pin count and high density packages that many processors now use such as quad flat packs, ball grid arrays and so on, the job of getting sockets that can reliably provide good electrical contacts is becoming harder. This is starting to restrict the ability of probe manufacturers to provide headers that will fit these sockets, assuming that the sockets are available in the first place.

  The ability to get several hundred individual signal cables into the probe is also causing problems and this has meant that for some processors, emulators are no longer a practical proposition.

- Matching the electrical characteristics

  This follows on from the previous point. The electrical characteristics of the probe should match that of the device the emulator is emulating. This includes the electrical characteristics of the pins. The difficulty is that the probe and its associated wiring make this matching very difficult indeed and in some cases, this imposes speed limits on the emulation or forces the insertion of wait states. Either way, the emulation is far from perfect and this can cause restrictions in the use of emulation. In some cases, where speed is of the essence, emulation can prevent the system from working at the correct design speed.

• Field servicing

This is an important but oft neglected point. It is extremely useful for a field engineer to have some form of debug access to a system in the field to help with fault identification and rectification. If this relies on an emulator, this can pose problems of access and even power supplies if the system is remote.

So, faced with these difficulties, many of the more recent processors have adopted different strategies to provide emulation support without having to resort to the traditional emulator and its inherent problems.

The basic methodology is to add some debugging support to the processor that enables a processor to be single stepped and breakpointed under remote control from a workstation or host. This facility is made possible through the provision of dedicated debug ports.

JTAG

JTAG ports were originally designed and standardised to provide a way of taking over the pins of a device to allow different bit patterns to be imposed on the pins to allow other devices within the circuit to be tested. This is important to implement boundary scan techniques without having to remove a processor. It allows access to all the hardware within the system.

The system works by using a serial port and clocking data into a large shift register inside the device. The outputs from the shift register are then used to drive the pins under control from the port.

OnCE

OnCE or on-chip emulation is a debug facility used on Motorola's DSP 56x0x family of DSP chips. It uses a special serial port to access additional registers within the device that provide control over the processor and access to its internal registers. The advantage of this approach is that by bringing out the OnCE port to an external connector, every system can provide its own in circuit emulation facilities by hooking this port to an interface port in a PC or workstation. The OnCE port allows code to be downloaded and single stepped, breakpoints to be set and the display of the internal registers, even while operating. In some cases, small trace buffers are available to capture key events.

BDM

BDM or background debug mode is provided on Motorola's MC683xx series of processors as well as some of the newer 8 bit microcontrollers such as the MC68HC12. It is similar in concept to OnCE, in that it provides remote control and access over the processor, but the way that it is done is slightly different. The processor has additional circuitry added which provides a special background debug mode where the processor does not execute any code but is under the control of the remote system connected to its BDM port. The BDM state is entered by the assertion of a BDM signal or by executing a special BDM instruction. Once the BDM mode has been entered, low level microcode takes over the processor and allows breakpoints to be set, registers to be accessed and single stepping to take place and so on, under command from the remote host.

10   Buffering and other data structures

This chapter covers the aspects of possibly the most used data structure within embedded system software. The use and understanding behind buffer structures is an important issue and can greatly effect the design of software.

What is a buffer?

A buffer is as its name suggests an area of memory that is used to store data usually on a temporary basis prior to processing it. It is used to compensate for timing problems between software modules or subsystems that cannot always guarantee to process every piece of data as it becomes available. It is also used as a collection point for data so that all the relevant information can be collected and organised before processing.

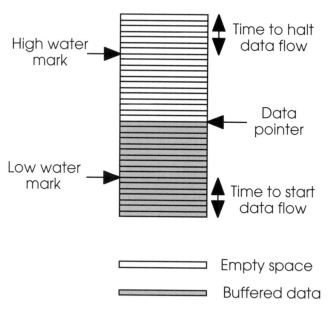

A basic buffer structure

The diagram shows the basic construction of a buffer. It consists of a block of memory and a pointer that is used to locate the next piece of data to be accessed or removed from the buffer. There are additional pointers which are used to control the buffer and prevent data overrun and underrun. An overrun occurs when the buffer cannot accept any more data. An underrun is caused when it is asked for data and cannot provide it.

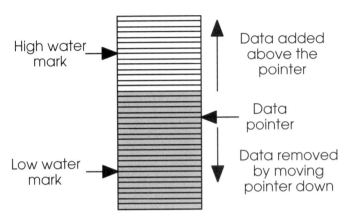

High water mark

Data added above the pointer

Data pointer

Low water mark

Data removed by moving pointer down

Adding and removing data

Data is removed by using the pointer to locate the next value and moving the data from the buffer. The pointer is then moved to the next location by incrementing its value by the number of bytes or words that have been taken. One common programming mistake is to confuse words and bytes. A 32 bit processor may access a 32 bit word and therefore it would be logical to think that the pointer is incremented by one. The addressing scheme may use bytes and therefore the correct increment is four. Adding data is the opposite procedure. The details on exactly how these procedures work determine the buffer type and its characteristics and are explained later in this chapter.

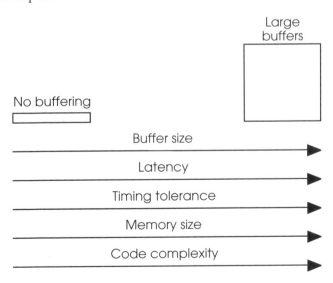

Large buffers

No buffering

Buffer size

Latency

Timing tolerance

Memory size

Code complexity

Buffering trade-offs

However, while buffering does undoubtedly offer benefits, they are not all for free and their use can cause problems. The diagram shows the common trade-offs that are normally encountered with buffering.

Latency

If data is stored in a buffer, then there is normally a delay before the first and subsequent data is received from the buffer. This delay is known as the buffer latency and in some systems, it can be a big advantage. In others, however, its effect is the opposite.

For a real-time system, the buffer latency defines the earliest that information can be processed and therefore any response to that information will be delayed by the latency irrespective of how fast or efficient the processing software and hardware is. If data is buffered so that eight samples must be received before the first is processed, the real-time response is now eight times the data rate for a single sample plus the processing time. If the first sample was a request to stop a machine or ignore the next set of data, the processing that determines its meaning would occur after the event it was associated with. In this case, the data that should have been ignored is now in the buffer and has to be removed.

Latency can also be a big problem for data streams that rely on real-time to retain their characteristics. For example, digital audio requires a consistent and regular stream of data to ensure accurate reproduction. Without this, the audio is distorted and can become unintelligible in the case of speech. Buffering can help by effectively having some samples in reserve so that the audio data is always there for decoding or processing. This is fine except that there is now an initial delay while the buffer fills up. This delay means an interaction with the stream is difficult as anyone who has had an international call over a satellite link with the large amount of delay can vouch for. In addition some systems cannot tolerate delay. Digital telephone handsets have to demonstrate a very small delay in the audio processing path which limits the size of any buffering for the digital audio data to less than four samples. Any higher and the delay caused by buffer latency means that the phone will fail its type approval.

Timing tolerance

Latency is not all bad, however, and used in the right amounts can provide a system that is more tolerant and resilient than one that is not. The issue is based around how time critical the system is and perhaps more importantly how deterministic is it.

Consider a system where audio is digitally sampled, filtered and stored. The sampling is performed on a regular basis and the filtering takes less time than the interval between samples. In this case, it is possible to build a system that does not need buffering and will have a very low latency. As each sample is received, it is processed and stored. The latency is the time to take a single sample.

If the system has other activities and especially if those involve asynchronous events such as the user pressing a button on the panel, then the guarantee that all the processing can be completed between samples may no longer be true. If this deadline is not made, then a sample may be lost. One solution to this — there are others such as using a priority system as supplied by a real-time operating system — is to use a small buffer to temporarily store the data so that it is not lost. By doing this the time constraints on the processing software are reduced and are more tolerant of other events. This is, however, at the expense of a slightly increased latency.

Memory size

One of the concerns with buffers is the memory space that they can take. With a large system this is not necessarily a problem but with a microcontroller or a DSP with on-chip memory, this can be an issue when only small amounts of RAM are available.

Code complexity

There is one other issue concerned with buffers and buffering technique and that concerns the complexity of the control structures needed to manage them. There is a definite trade-off between the control structure and the efficiency that the buffer can offer in terms of memory utilisation. This is potentially more important in the region of control and interfacing with interrupts and other real-time events. For example, a buffer can be created with a simple area of memory and a single pointer. This is how the frequently used stack is created. The control associated with the memory — or buffer which is what the memory really represents — is a simple register acting as an address pointer. The additional control that is needed to remember the stacking order and the frame size and organisation is built into the frame itself and is controlled by the microprocessor hardware. This additional level of control must be replicated either in the buffer control software or by the tasks that use the buffer. If a single task is associated with a buffer, it is straightforward to allow the task to implement the control. If several tasks use the same buffer, then the control has to cope with multiple and, possibly, conflicting accesses and while this can be done by the tasks, it is better to nominate a single entity to control the buffer. However, the code complexity associated with the buffer has now increased.

The code complexity is also dependent on how the buffer is organised. It is common for multiple pointers to be used along with other data such as the number of bytes stored and so on. The next section in this chapter will explain the commonly used buffer structures.

Linear buffers

The term linear buffer is a generic reference to many buffers that are created with a single piece of linear contiguous memory that is controlled by pointers whose address increments linearly. The examples so far discussed in this chapter are all examples of linear buffers.

The main point about them is that they will lose data when full and fail to provide data when empty. This is obvious but as will be shown, the way in which this happens with linear buffers compared to circular ones is different. With a linear buffer, it loses incoming data when full so that the data it does contain becomes older and older. This is the overrun condition. When it is empty, it will provide old data, usually the last entry, and so the processor will continue to process potentially incorrect data. This is the underrun condition.

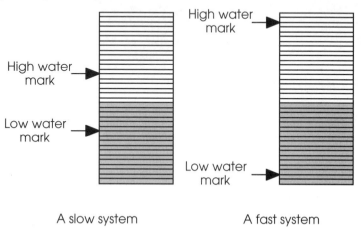

A slow system   A fast system

Adjusting the water marks

Within a real-time system, these conditions are often but not always considered error conditions. In some cases, the loss of data is not critical but with any data processing that is based on regular sampling, it will introduce errors. There are further complications concerning how these conditions are prevented from occurring. The solution is to use a technique where the pointers are checked against certain values and the results used to trigger an action such as fetching more data and so on. These values are commonly referred to as high and low water marks. This naming is that they are similar to the high and low water marks seen at the coast that indicate the minimum and maximum levels that tidal water will fall and rise.

The number of entries below the low water mark determine how many entries the buffer still has and thus the amount of time that is available to start filling the buffer before the buffer empties and the underrun condition exists. The number of empty entries in the buffer above the high water mark determines the length of time that is available to stop the further filling of the buffer and thus prevent data loss through overrun. By comparing the various input and output pointers with these values, events can be generated to start or stop filling the buffer. This could simply take the form of jumping to a subroutine, generating a software interrupt or within the context of an operating system posting a message to another task to fill the buffer.

Directional buffers

If you sit down and start playing with buffers, it quickly becomes more apparent that there is more to buffer design than first meets the eye. For example, the data must be kept in the same order in which it was placed to preserve the chronological order. This is especially important for signal data or data that is sampled periodically. With an infinitely long buffer, this is not a problem. The first data is placed at the top of the buffer and new data is inserted underneath. The data in and out pointers then simply move up and down as needed. The order is preserved because there is always space under the existing data entries for more information. Unfortunately, such buffers are rarely practical and problems can occur when the end of the buffer is reached. The previous paragraphs have described how water marks can be used to trigger when these events are approaching and thus give some time to resolve the situation.

The resolution is different depending on how the buffer is used, i.e. is it being used for inserting data, extracting data or both. The solutions are varied and will lead onto the idea of multiple buffers and buffer exchange. The first case to consider is when a buffer is used to extract and insert data at the same time.

Single buffer implementation

In this case the buffer is used by two software tasks or routines to insert or extract information. The problem with water marks is that they have data above or below them but the free space that is used to fill the buffer does not lie in the correct location to preserve the order.

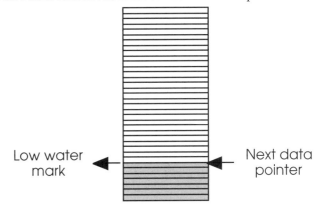

```
If next_data_pointer = low_water
        then
                copy current samples to top of buffer
                set next_data_pointer to top of buffer
                fill rest of buffer with new samples
        endif
```

Single buffer implementation

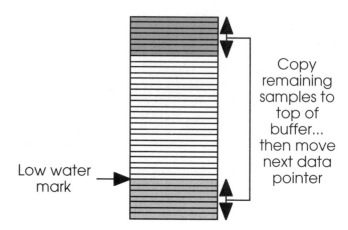

Low water
mark

Copy
remaining
samples to
top of
buffer...
then move
next data
pointer

Moving the samples and pointer

One solution to this problem is to copy the data to a new location and then continue to back-fill the buffer. This is the method shown in the next three diagrams.

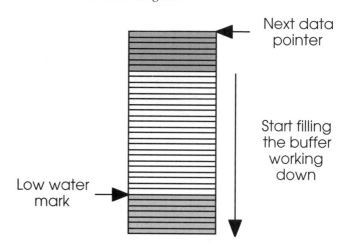

Next data
pointer

Start filling
the buffer
working
down

Low water
mark

Back-filling the buffer

It uses a single low water mark and a next data pointer. The next data pointer is used to access the next entry that should be extracted. Data is not inserted into the buffer until the next data pointer hits the low water mark. When that happens, the data below the low water mark is copied to the top of the buffer and the next data pointer moved to the top of the buffer. A new pointer is then created whose initial value is that of the first data entry below the copied data. This is chronologically the correct location and thus the buffer can be filled by using this new pointer. The original data at the bottom of the buffer can be safely overwritten because the data was copied to the top of the buffer. Data can still be extracted by the next data pointer.

When the temporary pointer reaches the end of the buffer, it stops filling. The low water mark — or even a different one — can be used to send a message to warn that filling must stop soon. By adding more pointers, it is possible to not completely fill the area below the low water mark and then use this to calculate the number of entries to move and thus the next filling location.

This method has a problem in that there is a delay while the data is copied. A more efficient alternative to copying the data, is to copy the pointer. This approach works by still using the low water mark, except that the remaining data is not copied. The filling will start at the top of the buffer and the next data pointer is moved to the top of the buffer when it hits the end. The advantage that this offers is that the data is not copied and only a pointer value is changed.

Both approaches allow simultaneous filling and extraction. However, care must be taken to ensure that the filling does not overwrite the remaining entries at the bottom of the buffer with the pointer copying technique, and that extracting does not access more data than has been filled. Additional pointer checking may be needed to ensure this integrity in all circumstances and not leave the integrity dependent on the dynamics of the system, i.e. assuming that the filling/extracting times will not cause a problem.

Double buffering

The problem with single buffering is that there is a tremendous overhead in managing the buffer in terms of maintaining pointers, checking the pointers against water marks and so on. It would be a lot easier to separate the filling from the extraction. It removes many of the contention checks that are needed and greatly simplifies the design. This is the idea behind double buffering.

Instead of a single buffer, two buffers are used with one allocated for filling and the second for extraction. The process works by filling the first buffer and passing a pointer to it to the extraction task or routine. This filled buffer is then simply used by the software to extract the data. While this is going on, the second buffer is filled so that when the first buffer is emptied, the second buffer will be full with the next set of data. This is then passed to the extraction software by passing the pointer. Many designs will recycle the first buffer by filling it while the second buffer is emptied. The process will add delay into system which will depend on the time taken to fill the first buffer.

Care must be taken with the system to ensure that the buffer swap is performed correctly. In some cases, this can be done by passing the buffer pointer in the time period between filling the last entry and getting the next one. In others, water marks can be used to start the process earlier so that the extraction task may be passed to the second buffer pointer before it is completely filled. This allows it the option of accessing data in the buffer if needed instead of having to wait for the buffer to complete filling. This is useful when the extraction timing is not consistent and/or requires different amounts of data. Instead of making the buffers the size of the largest data structure, they can be smaller and the double buffering used to ensure that data can be supplied. In other words, the double buffering is used to give the appearance of the presence of a far bigger buffer than is really there.

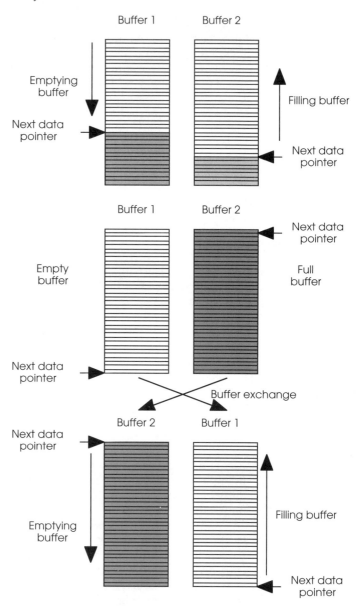

Double buffering

Buffer exchange

Buffer exchange is a technique that is used to simplify the control code and allow multiple tasks to process data simultaneously without having to have control structures to supervise access. In many ways it is a variation of the double buffering technique.

This type of mechanism is common to the SPOX operating system used for DSP processors and in these types of embedded systems it is relatively simple to implement.

The main idea of the system is the concept of exchanging empty buffers for full ones. Such a system will have at least two buffers although many more may be used. Instead of normally using a read or write operation where the data to be used during the transfer is passed as a parameter, a pointer is sent across that points to a buffer. This buffer would contain the data to be transferred in the case of a write or simply be an empty buffer in the case of a read. The command is handled by a device driver which returns another pointer that points to a second buffer. This buffer would contain data with a read or be empty with a write. In effect what happens is that a buffer is passed to the driver and another received back in return. With a read, an empty buffer is passed and a buffer full of data is returned. With a write, a full buffer is passed and an empty one is received. It is important to note that the buffers are different and that the driver does not take the passed buffer, use it and then send it back.

The advantages that this process offers:

- The data is not copied between the device driver and the requesting task.

- Both the device driver and the requesting task have their own separate buffer area and there is thus no need to have semaphores to control any shared buffers or memory.

- The requesting task can use multiple buffers to assimilate large amounts of data before processing.

- The device driver can be very simple to write.

- The level of inter-task communication to indicate when buffers are full or ready for collection can be varied and thus be designed to fit the end application or system.

There are some disadvantages however:

- There is a latency introduced dependent on the size of the buffer in the worst case. Partial filling can be used to reduce this if needed but requires some additional control to signify the end of valid data within a buffer.

- Many implementations assume a fixed buffer size which is predetermined, usually during the software compilation and build process. This has to be big enough for the largest message but may therefore be very inefficient in terms of memory usage for small and simple data. Variable size buffers are a solution to this but require more complex control to handle the length of the valid data. The buffer size must still be large enough for the biggest message and thus the problem of buffer size granularity may come back again.

• The buffers must be accessible by both the driver and requesting tasks. This may seem to be very obvious but if the device driver is running in supervisor mode and the requesting task is in the user mode, the memory management unit or address decode hardware may prevent the correct access. This problem can also occur with segmented architectures like the 8086 where the buffers are in different segments.

Linked lists

Linked lists are a way of combining buffers in a regular and methodical way using pointers to point to the next entry in the list. The linking is maintained by adding an entry to a buffer which contains the address to the next buffer or entry in the list. Typically, other information such as buffer size may be included as well as allowing the list to support different size entries. Each buffer will also include information for its control such as next data pointers and water marks depending on their design or construction.

With a single linked list, the first entry will use the pointer entry to point to the location of the second entry and so on. The last entry will have a special value which indicates that the entry is the last one.

New entries are added by breaking the link, extracting the link information and then inserting the new entry and remaking the link. When appending the entry to the end, the new entry pointer takes the value of the original last entry — the end of list special value in this case. The original last entry will update its link to point to the new entry and in this way the link is created.

The process for inserting in the middle of the list is shown in the diagram and follows the basic principles.

Linked lists have some interesting properties. It is possible to follow all the links down to a particular insertion point. Please note that this can only be done from the first entry down without storing additional information. With a single linked list like the one shown, there is no information in each entry to show where the link came from, only where it goes to. This can be overcome by storing the link information as you traverse down the list but this means that any search has to start at the top and work through, which can be a tiresome process.

The double linked list solves this by using two links. The original link is used to move down and the new link contains the missing information to move back up. This provides a lot more flexibility and efficiency when searching down the list to determine an entry point. If a list is simply used to concatenate buffers or structures together, then the single link list is more than adequate. If the ability to search up and down the list to reorder and sort the list or find different insertion points, then the double linked list is a better choice to consider.

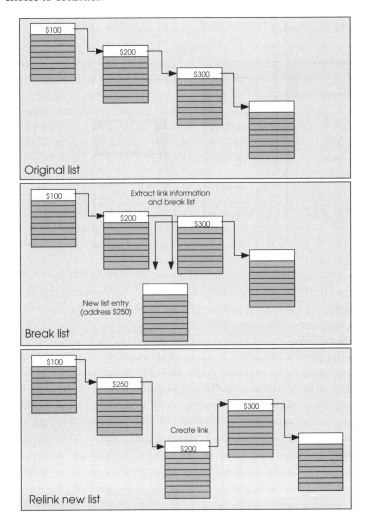

Inserting a new entry

FIFOs

FIFOs or first in, first out are a special form of buffer that uses memory to form an orderly queue to hold information. Its most important attribute is that the data can be extracted in the same way as it was entered. These are used frequently within serial comms chips to hold data temporarily while the processor is busy and cannot immediately service the peripheral. Instead of losing data, it is placed in the FIFO and extracted later. Many of the buffers described so far use a FIFO architecture.

Their implantation can be done either in software or more commonly with special memory chips that automatically maintain the pointers that are needed to control and order the information.

Circular buffers

Circular buffers are a special type of buffer where the data is circulated around a buffer. In this way they are similar to a single buffer that moves the next data pointer to the start of the buffer to access the next data. In this way the address pointer circulates around the addresses. In that case, care was taken so that no data was lost. It is possible to uses such a buffer and lose data to provide a different type of buffer structure. This is known as a circular buffer where the input data is allowed to overwrite the last data entries. This keeps the most recent data at the expense of losing some of the older data. This is useful for capturing trace data where a trace buffer may be used to hold the last n data samples where n is the size of the buffer. By doing this, the buffer updating can be frozen at any point and the last n samples can be captured for examination. This technique is frequently used to create trace and history buffers for test equipment and data loggers.

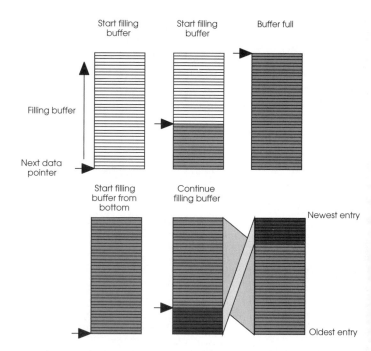

Circular buffers

This circular structure is also a very good match for coefficient tables used in digital signal processing where the algorithm iterates around various tables and performing arithmetic operations to perform the algorithm.

The only problem is that the next data pointer must be checked to know when to reset it to the beginning of the buffer. This provides some additional overhead. Many DSPs provide a special modulo addressing mode that will automatically reset the address for a particular buffer size. This buffer size is normally restricted to a power of two.

Allocating buffer memory

Buffer memory can be allocated in one of two generic ways: statically at build time by allocating memory through the linker and dynamically during run time by calling an operating system to allocate/program memory access.

Static allocation requires the memory needs to be defined before building the application and allocating memory either through special directives at the assembler level or through the definition of global, static or local variables within the various tasks within an application. This essential declares variables which in turn allocate storage space for the task to use. The amount of storage that is allocated depends on the variable type and definition. Strings and character arrays are commonly used.

malloc()

malloc() and its counterpart unmalloc() are system calls that were originally used within UNIX environments to dynamically allocate memory and return it. Their popularity has meant that these calls are supported in many real-time operating systems. The call works by passing parameters such as memory size and type, starting address and access parameters for itself and other tasks that need to access the buffer or memory that will be returned. This is a common programming error. In reply, the calling task receives a pointer that points at the start of the memory if the call was successful, or an error code if it was not. Very often, the call is extended to support partial allocation where the system will still return a pointer to the start of the memory along with an error/status message stating that not all the memory is available and that the x bytes were allocated. Some other systems would class this partial allocation as an error and return an error message indicating failure.

Memory that was allocated via malloc() calls can be returned to the memory pool by using the unmalloc() call along with the appropriate pointer. This allows the memory to be recycled and used/allocated to other tasks. This recycling allows memory to be conserved but at the expense of processing the malloc() and unmalloc() calls. Some software find this overhead is unacceptable and allocate memory statically at build time. It should be noted that in many cases, the required memory strategy may require a different design strategy. For memory efficient designs, requesting and recycling memory may be the best option. For speed where the recycling overhead cannot be tolerated, static allocation may be the best policy.

Memory leakage

Memory leakage is a term that is used to describe a bug that gradually uses all the memory within a system until such point that a request to use or access memory that should succeed fails. The term leakage is analogous to a leaking bucket where the contents gradually disappear. The contents within an embedded system are memory.

The common symptoms are stack frame errors caused by the stack overflowing its allocated memory space and malloc() or similar calls to get memory failing. There are several common programming faults that cause this problem.

Stack frame errors

It is common within real-time systems, especially those with nested exceptions, to use the exception handler to clean up the stack before returning to the previous executing software thread or to a generic handler. The exception context information is typically stored on the stack either automatically or as part of the initial exception routine. If the exception is caused by an error, then there is probably little need to return execution to the point where the error occurred. The stack, however, contains a frame with all this return information and therefore the frames need to be removed by adjusting the stack pointer accordingly. It is normally this adjustment where the memory leakage occurs.

- Adjusting the stack for the wrong size frame. If the adjustment is too large, then other stack frames can be corrupted. If it is too small, then at best some stack memory has been lost and at worst the previous frame can be corrupted.

- Adjusting the stack pointer by the wrong value, e.g. using the number of words in the frame instead of the number of bytes.

- Setting the stack pointer to the wrong address so that it is on an odd byte boundary, for example.

Failure to return memory to the memory pool

This is a common cause of bus and memory errors. It is caused by tasks requesting memory and then not releasing it when their need for it is over. It is good practice to ensure that when a routine uses malloc() to request memory that it also uses unmalloc() to return it and make it available for reuse. If a system has been designed with this in mind, then there are two potential scenarios that can occur that will result in a memory problem. The first is that the memory is not returned and therefore subsequent malloc() requests cannot be serviced when they should be. The second is similar but may only occur in certain circumstances. Both are nearly always caused by failure to return memory when it is finished, but the error may not occur until far later in time. It may be the same task asking for memory or another that causes the problem to arise. As a result, it can be difficult to detect which task did not return the memory and is responsible for the problem.

In some cases where the task may return the memory at many different exit points within its code — this could be deemed as a bad programming practice and it would be better to use a single exit sub-routine for example — it is often a programming omission at one of these points that stops the memory recycling.

It is difficult to identify when and where memory is allocated unless some form of record is obtained. With memory management systems, this can be derived from the descriptor tables and address translation cache entries and so on. These can be difficult to retrieve and decode and so a simple transaction record of all malloc() and unmalloc() calls along with a time stamp can prove invaluable. This code can be enabled for debugging if needed by passing a DEBUG flag to the preprocessor. Only if the flag is true will it compile the code.

Housekeeping errors

- Access rights not given

  This is where a buffer is shared between tasks and only one has the right access permission. The pointer may be passed correctly by using a mailbox or message but any access would result in a protection fault or if mapped incorrectly in accessing the wrong memory location.

- Pointer corruption

  It is very easy to get pointers mixed up and thus use or update the wrong one and thus corrupt the buffer.

- Timing problems with high and low water marks

  Water marks are used to provide early warning and should be adjusted to cope with worst case timings. If not, then despite their presence, it is possible to get data overrun and underrun errors.

11   Memory and performance trade-offs

This chapter describes the trade-offs made when designing an embedded system to cope with the speed and performance of the processor in doing its tasks. The problem faced by many designers is that the overall design requires a certain performance level in terms of processing or data throughput which on first appearance is satisfied by a processor. However, when the system is actually implemented, its performance is lacking due to the performance degradation that the processor can suffer as a result of its memory architecture, its I/O facilities and even the structure and coding techniques used to create the software.

The effect of memory wait states

This is probably the most important factor to consider as it can have the biggest impact on the performance of any system. With most high performance CPUs such as RISC and DSP processors offering single cycle performance where one or more instructions are executed on every clock edge, it is important to remember the conditions under which this is permitted:

- Instruction access is wait state free

  To achieve this, the instructions are fetched either from internal on-chip memory (usually wait state free but not always), or from internal caches. The problem with caches is that they only work with loops so that the first time through the loop, there is a performance impact while the instructions are fetched from external memory. Once the instructions have been fetched, they are then available for any further execution and it is here that the performance starts to improve.

- Data access is wait state free

  If an instruction manipulates data, then the time taken to execute the instruction must include the time needed to store the results of any data manipulation or access data from memory or a peripheral I/O port. Again, if the processor has to wait — known as stalling — while data is stored or read, then performance is lost. If an instruction modifies some data and it takes five clocks to store the result, this potentially can cause processing power to be lost. In many cases, processor architectures make the assumption that there is wait state-free data access by either using local memory, registers or cache memory to hold the information.

- There are no data dependencies outstanding.

  This leads on from the previous discussion and concerns the ability of an instruction to immediately use the result from a previous instruction. In many cases, this is only permitted if there is a delay to allow the processor to synchronise itself. As a result, the single cycle delay has the same result as a single cycle wait state and thus the performance is degraded.

As a result of all these conditions, it should not be assumed that a 80 MHz single cycle instruction processor such as a DSP- or a RISC-based machine can provide 80 MIPs of processing power. It can provided the conditions are correct and there are no wait states, data dependencies and so on. If there are, then the performance must be degraded. This problem is not unrecognised and many DSP and processor architectures utilise a lot of silicon in providing clever mechanisms to reduce the performance impact. However, the next question that must be answered is how do you determine the performance degradation and how can you design the code to use these features to minimise any delay?

Scenario 1 — Single cycle processor with large external memory

In this example, there is a single cycle processor that has to process a large external table by reading a value, processing it and writing it back. While the processing level is small — essentially a data fetch, three processing instructions and a data store — the resulting execution time can be almost three times longer than expected. The reason is stalling due to data latency. The first figure shows how the problem can arise.

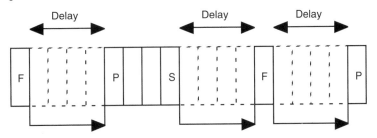

At 100 MHz clock:
Theoretical time for 5 instructions = 50 ns
Practical time for 5 instructions = 130 ns

Stalling due to data latency

The instruction sequence consists of a data fetch followed by three instructions that process the information before storing the result. The data access goes to external memory where there are access delays and therefore the first processing instruction must wait until the data is available. In this case, the instruction execution is stalled and thus the fetch instruction takes the equivalent of five cycles instead of the expected single cycle. Once the data has been received, it can be processed and the processing instructions will continue, one per clock. The final instruction stores the end result and again this is where further delays can be experienced. The store instruction experiences the same external memory delays as the fetch instruction. However, its data is not required by the next set of instructions and therefore the rest of the instruction stream should be able to continue. This is not the case. The next instruction is a fetch and has to compete with the external bus interface which is still in use by the preceding store. As a result, it must also wait until the transaction is completed.

The next processing instruction now cannot start until the second fetch instruction is completed. These delays mean that the total time taken at 100 MHz for the 5 instructions (1 fetch + 3 processing + 1 store) is not 50 ns but 130 ns — an increase of 2.6 times.

The solution to this involves reordering the code so that the delays are minimised by overlapping operations. This assumes that the processor can do this, i.e. the instructions are stored in a separate memory space that can be accessed simultaneously with data. If not, then this conflict can create further delays and processor stalls. The basic technique involves moving the processing segment of the code away from the data access so that the delays do not cause processing stalls because the data dependencies have been removed. In other words, the data is already available before the processing instructions need it.

Moving dependencies can be achieved by restructuring the code so that the data fetch preceding the processing fetches the data for the next processing sequence and not the one that immediately follows it.

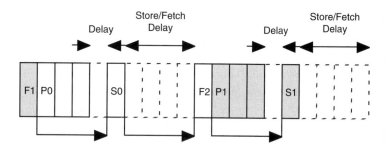

At 100 MHz clock:
Theoretical time for 5 instructions =   50 ns
Practical time for 5 instructions   = 100 ns

Removing stalling

The diagram above shows the general approach. The fetch instruction is followed by the processing and storage instruction for the preceding fetch. This involves using an extra register or other local storage to hold the sample until it is needed but it removes the data dependency. The processing instructions P0 onward that follow the fetch instruction F1 do not have any data dependency and thus can carry on processing. The storage instruction S0 has to wait one cycle until F0 has completed and similarly the fetch instruction F2 must wait until S0 has finished. These delays are still there because of the common resource that the store and fetch instructions use, i.e. the external memory interface. By reordering in this way, the five instruction sequence is completed twice in every 20 clocks giving a 100 ns timing which is a significant improvement.

This example also shows that the task in this case is I/O bound in that the main delays are caused by waiting for data to be fetched or stored. If the processing load could almost be doubled and further interleaved with the store operations without changing or delaying the data throughput of the system. What would happen, however, is an increase in the processing load that the system could handle.

The delays that have been seen are frequently exploited by optimising routines within many modern compilers. These compilers know from information about the target processor when these types of delays can occur and how to reschedule instructions to make use of them and regain some of the lost performance back.

Scenario 2 — Reducing the cost of memory access

The preceding scenario shows the delays that can be caused by accessing external memory. If the data is accessible from a local register the delay and thus the performance loss is greatly reduced and may be zero. If the data is in local on-chip memory or in an on-chip cache, the delay may only be a single cycle. If it is external DRAM, the delay may be 9 or 10 cycles. This demonstrates that the location of data can have a dramatic effect on any access delay and the resultant performance loss.

The good way of tackling this problem is to create a table with the storage location, its storage capability and speed of access in terms of clock cycles and develop techniques to move data between the various locations so that it is available when the processor needs it. For example, moving the data into registers compared to direct manipulation externally in memory can reduce the number of cycles needed, even when the load of saving and restoring the registers contents to free up the storage is taken into account.

Using registers

Registers are the fastest access storage resource available to the processor and are the smallest in size. As a result they are an extremely scarce resource which has to be used and managed carefully. The main problem is in deciding what information to store and when. This dilemma comes from the fact that there is frequently insufficient register space to store all the data all of the time. As a result, registers are used to store temporary values before updating main memory with a final result, to hold counter values for loop constructions and so on and for key important values. There are several approaches to doing this and exploiting their speed:

- Let the compiler do the work

  Many compilers will take care of register management automatically for you when it is told to use optimisation techniques. For many applications that are written in a high level language such as C, this is often a good choice.

- Load and store to provide faster local storage

  In this case, variables stored in external memory or on a stack are copied to an internal register, processed and then the result is written back out. With RISC processors that do not support direct memory manipulation, this is the only method of manipulating data. With CISC processors, such as the M68000 family, there is a choice. By writing code so that data is brought in to be manipulated, instead of using addressing modes that operate directly on external memory, the impact of slow memory access can be minimised.

- Declaring register-based variables

  By assigning a variable to a register during its initial declaration, the physical access to the variable will be quicker. This can be done explicitly or implicitly. Explicit declarations use special attributes that the programmer uses in the declaration, e.g. reg. An implicit declaration is where the compiler will take a standard declaration such as global or static and implicitly use this to allocate a register if possible.

- Context save and restore

  If more variables are assigned to registers than there are registers within the processor, the software may need to perform a full save and restore of the register set before using or accessing a register to ensure that these variables are not corrupted. This is an accepted process when multitasking so that the data of one task that resides in the processor registers is not corrupted. This procedure may need to be used at a far lower level in the software to prevent inadvertent register-based data corruption.

Using caches

Data and instruction caches work on the principle that both data and code is accessed more than once. The cache memory will store the information as it is fetched from the main memory so that any subsequent access is from the faster cache memory. This assumption is important because straight line code without branches, loops and so on will not benefit from a cache.

The size and organisation of the cache is important because it determines the efficiency of the overall system. If the program loops will fit into the cache memory, the fastest performance will be realised and the whole of the external bus bandwidth will be available for data movement. If the data cache can hold all the data structures, then the data access will be the fastest possible. In practice, the overall performance gain is less than ideal because inevitably access to the external memory will be needed either to fetch the code and data the first time around, when the cache is not big enough to contain all the instructions or data, or when the external bus must be used to access an I/O port where data cannot be cached. Interrupts and other asynchronous events will also compete for the cache and can cause instructions and data that has been cached prior to the event to be removed, thus forcing external memory accesses when the original program flow is continued.

Preloading caches

One trick that can be used with caches is to preload them so that a cache miss is never encountered. With normal operation, a cache miss will force an external memory access and while this is in progress, the processor is stalled waiting for the information — data or instruction — to be returned. In many code sequences, this is more likely to happen with data, where the first time that the cache and external bus is used to access the data is when it is needed. As described earlier with scenario 1, this delay occurs at an important point in the sequence and the delay prevents the processor from continuing.

By using the same technique as used in scenario 1, the data cache can be preloaded with information for the next processing iteration before the current one has completed. The PowerPC architecture provides special instructions that allow this to be performed. In this way, the slow data access is overlapped with the processing and data access from the cache and does not cause any processor stalls. In other words, it ensures that the cache always continues to have the data that the instruction stream needs to have.

By the very nature of this technique, it is one that is normally performed by hand and not automatically available through the optimisation techniques supplied with high level language compilers.

It is very useful with large amounts of data that would not fit into a register. However, if the time taken to fetch the data is greater than the time taken to process the previous block, then the processing will stall.

Caches also have one other interesting characteristic in that they can make it very difficult to predict how long a particular operation will take to execute. If everything is in the cache, the time will be short. If not then it will be long. In practice, it will be somewhere in between. The problem is that the actual time will depend on the number of cache hits and misses which will depend in turn on the software that has run before which will have overwritten some of the entries. As a result, the actual timing becomes more statistical in nature and in many cases the worst case timing has to be assumed, even though statistically the routine will execute faster 99.999% of the time!

Using on-chip memory

Some microcontrollers and DSP chips have local memory which can be used to store data or instructions and thus offers fast local storage. Any access to it will be fast and thus data and code structures will always gain some benefit if they are located here. The problem is that to gain the best benefit, both the code and data structures must fit in the on-chip memory so that no external accesses are necessary. This may not be possible for some programs and therefore decisions have to be made on which parts of the code and data structures are allocated this resource. With a real-time operating system, local on-chip memory is often used to gain the best context switching time. This memory requirement now has to compete with algorithms that need on-chip storage to meet the performance requirements by minimising any processor stalls.

One good thing about using on-chip memory is that it makes performance calculations easier as the memory accesses will have a consistent access time.

Using DMA

Some microcontrollers and DSPs have on-chip DMA controllers which can be used in conjunction with local memory to create a sort of crude but efficient cache. In reality, it is more like a buffering technique with an intelligent agent filling up and controlling the buffers in parallel with the processing.

The basic technique defines the local memory into two or more buffers and programs the DMA controller to transfer data from the external memory to the local on-chip memory buffer while the data in the other buffer is processed. The overlapping of the DMA data transfer and the processing means that the data processing has local access to its data instead of having to wait for far slower memory access.

The buffering technique can be made more sophisticated by incorporating additional DMA transfers to move data out of the local memory back to the external memory. This may require the use of many more smaller buffers with different DMA characteristics. Constants could be put into one buffer which are read in but not read out. Variables can be stored in another where the information is written out to external memory.

Making the right decisions

The main problems faced by designers with these techniques is in knowing which one(s) should be used. The problem is that they involve a high degree of knowledge about the processor and the system characteristics. While a certain amount of information can be obtained from a documentation-based analysis, the use of simulation tools to run through code sequences and provide information concerning cache hits ratios, processor stalls and so on is a vital part in obtaining the optimum solution. Because of this, many cycle level processor simulation tools are becoming available which help provide this level of information.

12   Software examples

Benchmark example

The difficulty faced here appears to be a very simple one, yet actually poses an interesting challenge. The goal was to provide a simple method of testing system performance of different VMEbus processor boards to enable a suitable board to be selected. The problem was not how to measure the performance — there were plenty of benchmark routines available — but how to use the same compiler to create code that would run on several different systems with the minimum of modification. The idea was to generate some code which could then be tested on several different VMEbus target systems to obtain some relative performance figures. The reason for using the compiler was that typical C routines could be used to generate the test code.

The first decision made was to restrict the C compiler to non-I/O functions so that a replacement I/O library was not needed for each board. This still meant that arithmetic operations and so on could be performed but that the ubiquitous printf statements and disk access would not be supported. This decision was more to do with time constraints than anything else. Again for time reasons, it was decided to use the standard UNIX- based M680x0 cc compiler running on a UNIX system. The idea was not to test the compiler but to provide a vehicle for testing relative performance. Again, for this reason, no optimisation was done.

A simple C program was written to provide a test vehicle as shown. The exit() command was deliberately inserted to force the compiler to explicitly use this function. UNIX systems normally do not need this call and will insert the code automatically. This can cause difficulties when trying to examine the code to see how the compiler produces the code and what is needed to be modified.

```
main()
{
int a,b,c;

a=2;
b=4;
c=b-a;
b=a-c;
exit();
}
```

The example C program

The next stage was to look at the assembler output from the compiler. The output is different from the more normal M68000 assembler print out for two reasons. UNIX-based assemblers use different mnemonics compared to the standard M68000 ones and, secondly, the funny symbols are there to prompt the linker to fill in the addresses at a later stage.

The appropriate assembler source for each line is shown under the line numbers. The code for line 4 of the C source appears in the section headed ln 4 and so on. Examining the code shows that some space is created on the stack first using the link.l instruction. Lines 4 and 5 load the values 2 and 4 into the variable space on the stack. The next few instructions perform the subtraction before the jump to the exit subroutine.

This means that provided the main entry requirements are to set-up the stack pointer to a valid memory area, the code located at a valid memory address and the exit routine replaced with one more suitable for the target, the code should execute correctly. The first point can be solved during the code downloading. The other two require the use of the linker and replacement runtime routine for exit.

```
file      "math.c"
    data  1
    text
    def   main; val    main; scl   2;     type   044;  endef
    global        main
main:
    ln    1
    def   ~bf;  val    ~;     scl   101;  line   2;     endef
    link.l      %fp,&F%1
#movm.l &M%1,(4,%sp)
#fmovm  &FPM%1,(FPO%1,%sp)
    def   a;    val    -4+S%1;   scl   1;    type   04;    endef
    def   b;    val    -8+S%1;   scl   1;    type   04;    endef
    def   c;    val    -12+S%1;  scl   1;    type   04;    endef
    ln    4
    mov.l &2,((S%1-4).w,%fp)
    ln    5
    mov.l &4,((S%1-8).w,%fp)
    ln    6
    mov.l ((S%1-8).w,%fp),%d1
    sub.l     ((S%1-4).w,%fp),%d1
    mov.l %d1,((S%1-12).w,%fp)
    ln    7
    mov.l ((S%1-4).w,%fp),%d1
    sub.l     ((S%1-12).w,%fp),%d1
    mov.l %d1,((S%1-8).w,%fp)
    ln    8
    jsr   exit
L%12:
    def   ~ef;  val    ~;     scl   101;  line   9;     endef
    ln    9
#fmovm  (FPO%1,%sp),&FPM%1
#movm.l (4,%sp),&M%1
    unlk  %fp
    rts
    def   main; val    ~;     scl   -1;   endef
    set   S%1,0
    set   T%1,0
    set   F%1,-16
    set   FPO%1,4
    set   FPM%1,0x0000
    set   M%1,0x0000
    data  1
```

The resulting assembler source code

All the target boards have an onboard debugger which provides a set of I/O functions including a call to restart the debugger.

This would be an ideal way of terminating the program as it would give a definite visual signal of the termination of the software. So what was required was a routine that executed this debugger call. The routine for a Flight MC68020 evaluation board (EVM) is shown. This method is generic for M68000-based VMEbus boards. The other routines were very similar and basically used a different trap call number, e.g. TRAP #14 and TRAP #15 as opposed to TRAP #11. The global statement defines the label exit as an external reference so that the linker can recognise it. Note also the slightly different syntax used by the UNIX assembler. The byte storage command inserts zeros in the following long word to indicate that this is a call to restart the debugger.

```
exit:
global exit
    trap  &11
    byte  0,0,0,0
```

The exit() routine for the MC68020 EVM

This routine was then assembled into an object file and linked with the C source module using the linker. By including the new exit module on the command line with the C source module, it was used instead of the standard UNIX version. If this version was executed on the UNIX machine, it caused a core dump because a TRAP #11 system call is not normal.

```
SECTIONS
{
    GROUP 0x400600:
    {
            .text : { }
            .data : { }
            .bss  : { }
    }
}
```

The MC68020 EVM linker command file

The next issue was to relocate the code into the correct memory location. With a UNIX system, there are three sections that are used to store code and data, called .text, .data and .bss. Normally these are located serially starting at the address $00000000. UNIX with its memory management system will translate this address to a different physical address so that the code can execute correctly, instead of corrupting the M68000 vector table which is physically located at this address. With the target boards, this was not possible and the software had to be linked to a valid absolute address.

This was done by writing a small command file with SEC-TIONS and GROUP commands to instruct the linker to locate the software at a particular absolute address. The files for the MC68020 EVM and for the VMEbus board are shown. This file is included with the other modules on the command line.

```
SECTIONS
{
    GROUP 0x10000:
```

```
{
        .text :{}
        .data :{}
        .bss :{}
}
}
```

The VMEbus board linker command file

To download the files, the resulting executable files were converted to S-records and downloaded via a serial port to the respective target boards. Using the debugger, the stack pointer was correctly set to a valid area and the program counter set to the program starting address. This was obtained from the symbol table generated during the linking process. The program was then executed and on completion, returned neatly to the debugger prompt, thus allowing time measurements to be made. With the transfer technique established, all that was left was to replace the simple C program with more meaningful code.

To move this code to different M68000-based VMEbus processors is very simple and only the exit() routine with its TRAP instruction needs to be rewritten. To move it to other processors would require a change of compiler and a different version of the exit() routine to be written. By adding some additional code to pass and return parameters, the same basic technique can be extended to access the onboard debugger I/O routines to provide support for printf() statements and so on. Typically, replacement putchar() and getchar() routines are sufficient for terminal I/O.

Creating software state machines

With many real-time applications, a common approach taken with software is first to determine how the system must respond to external stimulus and then to create a set of state diagrams which correspond with these definitions. With a state diagram, a task or module can only exist in one of the states and can only make a transition to another state provided a suitable event has occurred. While these diagrams are easy to create, the software structure can be difficult.

One way of creating the equivalent of software state diagrams is to use a modular approach and message passing. Each function or state within the system — this can be part of or a whole state diagram — is assigned to a task. The code for the task is extremely simple in that it will do nothing and will wait for a message to arrive. Upon receipt of the message, it will decode it and use the data to change state. Once this has been completed, the task will go back to waiting for further input. The processing can involve other changes of state as well. Referring back to the example, the incoming interrupt will force the task to come out of its waiting state and read a particular register. Depending on the contents of that register, one of two further states can be taken and so on until the final action is to wait for another interrupt.

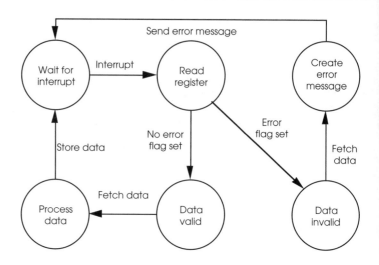

A state diagram

This type of code can be written without using an operating system, so what are the advantages? With a multitasking real-time operating system, other changes of state can happen in parallel. By allocating a task to each hardware interrupt, multiple interrupts can easily be handled in parallel. The programmer simply codes each task to handle the appropriate interrupt and the operating system takes care of running the multiple tasks. In addition, the operating system can easily resolve which interrupt or operation will get the highest priority. With complex systems, the priorities may need to change dynamically. This is an easy task for an operating system to handle and is easier to write compared to the difficulty of writing straight line code and coping with the different pathways through the software. The end result is easier code development, construction and maintenance.

The only interface to the operating system is in providing the message and scheduling system. Each task can use messages or semaphores to trigger its operation and during its processing, generate messages and toggle semaphores which will in turn trigger other tasks. The scheduling and time sharing of the tasks are also handled by the operating system.

In the example shown overleaf, there are six tasks with their associated mailboxes or semaphore. These interface to the real time operating system which handles message passing and semaphore control. Interrupts are routed from the hardware via the operating system, but the tasks can access registers, ports and buffers directly.

If the hardware generates an interrupt, the operating system will service it and then send a message to the sixth task to perform some action. In this case, it will read some registers. Once read, the task can now pass the data on to another task for processing. This is done via the operating system. The task supplies the data either directly or by using a memory pointer to it and the address of the receiving mail box. The operating system then places this message into the mail box and the receiving task is woken up. In reality, it is placed on the operating system scheduler ready list and allowed to execute.

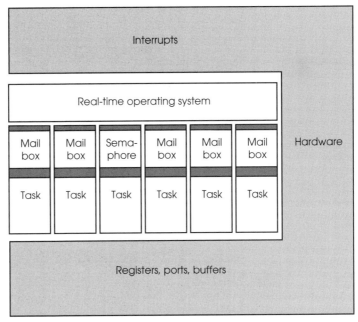

Basic system organisation

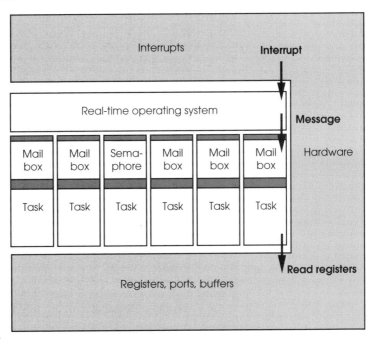

Handling an interrupt

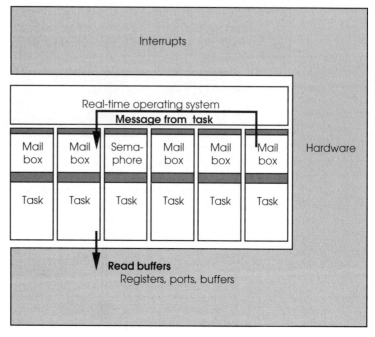

Handling a message from a task

Once woken up, the receiving task can then accept the message and process it. The message is usually designed to contain a code. This data may be an indication of a particular function that the task is needed to perform. Using this value, it can check to see if this function is valid given its current state and, if so, execute it. If not, the task can return an error message back via the operating system.

```
for_ever
    {
    Wait_for_ message();
    Process message();
    Discard_message();
    }
```

Example task software loop

Coding this is relatively easy and can be done using a simple skeleton program. The mechanism used to select the task's response is via two pointers. The first pointer reflects the current state of the task and points to an array of functions. The second pointer is derived from the message and used to index into the array of functions to execute the appropriate code. If the message is irrelevant, then the selected function may do nothing. Alternatively, it may process information, send a message to other tasks or even change its own current state pointer to select a different array of functions. This last action is synonymous to changing state.

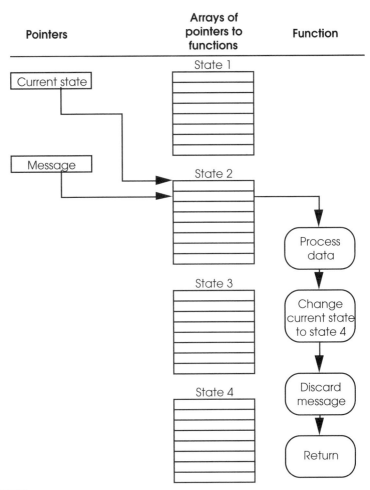

Responding to a message

Priority levels

So far in this example, the tasks have been assumed to have equal priority and that there is no single task that is particularly time critical and must be completed before a particular window expires. In real applications, this is rarely the case and certain routines or tasks are critical to meeting the system requirements. There are two basic ways of ensuring this depending on the facilities offered by the operating system. The first is to set the time critical tasks and routines as the highest priority. This will normally ensure that they will complete in preference to others. However, unless the operating system is pre-emptive and can halt the current task execution and swap it for the time critical one, a lower priority task can still continue execution until the end of its time slot. As a result, the time critical task may have to wait up to a whole time slice period before it can start. In such worse cases, this additional delay may be too long for the routine to complete in response to the original demand or interrupt. If the triggers are asynchronous,, i.e. can happen at any time and are not particularly tied to any one event, then the lack of pre-emption can cause a wide range of timings. Unfortunately for real-time systems, it is the worst case that has to be assumed and used in the design.

An alternative offered by some operating systems is the idea of an explicit lock where the task itself issues directives to lock itself into permanent execution. It will continue executing until it removes the lock. This is ideal for very time critical routines where the process cannot be interrupted by a physical interrupt or a higher priority task switch. The disadvantage is that it can lead to longer responses by other tasks and in some extreme cases system lock-ups when the task fails to remove the explicit lock. This can be done either with the technique of special interrupt service routines or through calls to the operating system to explicitly lock execution and mask out any other interrupts. Real-time operating systems usually offer at least one or other of these techniques.

Explicit locks

With this technique, the time critical software will make a system call prior to execution which will tell the operating system to stop it from being swapped out. This can also include masking out all maskable interrupts so that only the task itself or a non-maskable interrupt can interrupt the proceedings. The problem with this technique is that it greatly affects the performance of other tasks within the system and if the lock is not removed can cause the task to hog all of the processing time. In addition, it only works once the time critical routine has been entered. If it has to wait until another task has finished then the overall response time will be much lower.

Interrupt service routines

Some operating systems, such as pSOS⁺, offer the facility of direct interrupt service routines or ISRs where time critical code is executed directly. This allows critical software to execute before other priority tasks would switch out the routines as part of a context switch. It is effectively operating at a very low level and should not be confused with tasks that will activate or respond to a message, semaphore or event. In these cases, the operating system itself is working at the lower level and effectively supplies its own ISR which in turn passes messages, events and semaphores which activate other tasks.

The ISR can still call the operating system, but it will hold any task switching and other activities until the ISR routine has completed. This allows the ISR to complete without interruption.

It is possible for the ISR to send a message to its associated task to start performing other less time critical functions associated with the interrupt. If the task was responsible for reading data from a port, the ISR would read the data from the port and clear the interrupt and send a message to its task to process the data further. After completing, the task would be activated and effectively continue the processing started by the ISR. The only difference is that the ISR is operating at the highest possible priority level while the task can operate at whatever level the system demands.

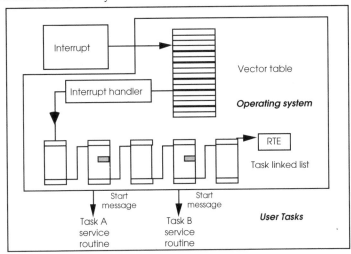

Handling interrupt routines within an
operating system

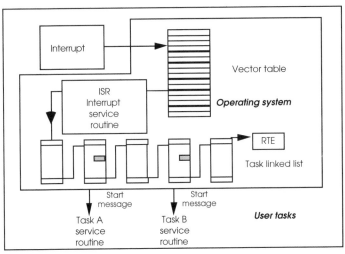

Handling interrupt routines using an ISR

Setting priorities

Given all these different ways of synchronising and controlling tasks, how do you decide which ones to use and how to set them up? There is no definitive answer to this as there are many solutions to the same problem, depending on the detailed characteristics that the system needs to exhibit. The best way to illustrate this is to take an example system and examine how it can be implemented.

The system shown in the diagram below consists of three main tasks. Task A receives incoming asynchronous data and passes this information onto task B which processes it. After processing, task C takes the data and transmits it synchronously as a packet. This operation by virtue of the processing and synchronous nature cannot be interrupted. Any incoming data can fortunately be ignored during this transmission.

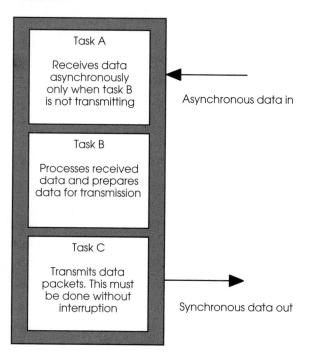

Example system

Task A highest priority

In this implementation, task priorities are used with A having the highest followed by C and finally B. The reasoning behind this is that although task C is the most time critical, the other tasks do not need to execute while it is running and therefore can simply wait until C has completed the synchronous transmission. When this has finished C can wake them up and make itself dormant until it is next required. Task A with its higher priority is then able to pre-empt B when an interrupt occurs, signalling the arrival of some incoming data.

However, this arrangement does require some careful consideration. If task A was woken up while C was transmitting data, task A would replace C by virtue of its higher priority. This would cause problems with the synchronous transmission. Task A could be woken up if it uses external interrupts to know when to receive the asynchronous data. So the interrupt level used by task A must be masked out or disabled prior to moving into a waiting mode and allowing task C to transfer data. This also means that task A should not be allocated a non-maskable interrupt.

Task C highest priority

An alternative organisation is to make task C the highest priority. In this case, the higher priority level will prevent task A from gaining any execution time and thus prevent any interrupt from interfering with the synchronous operation. This will work fine providing that task C is forced to be in a waiting mode until it is needed to transmit data. Once it has completed the data transfer, it would remove itself from the ready list and wait, thus allowing the other tasks execution time for their own work.

Using explicit locks

Task C would also be a candidate for using explicit locks to enable it to override any priority scheme and take as much execution time as necessary to complete the data transmission. The key to using explicit locks is to ensure that the lock is set by all entry points and is released on all exit points. If this is not done, the locks could be left on and thus lock out any other tasks and cause the system to hang up or crash.

Round robin

If a round robin system was used, then the choice of executing task would be chosen by the tasks themselves and this would allow task C to have as much time as it needed to transfer its data. The problem comes with deciding how to allocate time between the other two tasks. It could be possible for task A to receive blocks of data and then pass them onto task B for processing. This gives a serial processing line where data arrives with task A, is processed with task B and transmitted by task C. If the nature of the data flow matches this scenario and there is sufficient time between the arrival of data packets at task A to allow task B to process them, then there will be no problem. However, if the arriving data is spread out, then task B's execution may have to be interleaved with task A and this may be difficult to schedule, and be performed better by the operating system itself.

Using an ISR routine

In the scenarios considered so far, it has been assumed that task A does not overlap with task C and they are effectively mutually exclusive. What happens if this is not the case? It really depends on the data transfer rates of both tasks. If they are slow, i.e. the need to send or receive a character is slower than the context switching time, then the normal priority switching can be used with one of the tasks allocated the highest priority. With its synchronous communication, it is likely that it would be task C.

The mechanism would work as follows. Both tasks would be allocated their own interrupt level with C allocated the higher priority to match that of its higher task priority. This is important otherwise it may not get an opportunity to respond to the interrupt routine itself. Its higher level hardware interrupt may force the processor to respond faster but this good work could be negated if the operating system would not allocate a time slice because an existing higher priority task was already running.

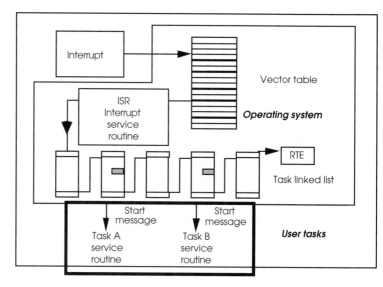

Responding to interrupts

If task A was servicing its own interrupt and a level C interrupt was generated, task A would be pre-empted, task C would start executing and on completion put itself back into waiting mode and thus allow task A to complete. With a pre-emptive system, the worst case latency for task C would be the time taken by the processor to recognise the external task C interrupt, the time taken by the operating system to service it and finally the time taken to perform a context switch. The worst case latency for task A is similar, but with the addition of the worst case value for task C plus another context switch time. The total value is the time taken by the processor to recognise the external task A interrupt, the time taken by the operating system to service it and, finally, the time taken to perform a context switch. The task C latency and context switch time must be added because the task A sequence could be interrupted and replaced at any time by task C. The extra context switch time must be included for when task C completes and execution is switched back to task A.

Provided these times and the time needed to execute the appropriate response to the interrupt fit between the time critical windows, the system will respond correctly. If not then time must be saved.

The diagram shows the mechanism that has been used so far which relies on a message to be sent that will wake up a task. It shows that this operation is at the end of a complex chain of events and that using an ISR, a lot of time can be saved.

The interrupt routines of tasks A and C would be defined as ISRs. These would not prevent context switches but will reduce the decision making overhead to an absolute minimum and is therefore more effective.

If the time windows still cannot be met, the only solution is to improve the performance of the processor or use a second processor dedicated to one of the I/O tasks.

13   Design examples

Burglar alarm system

This example describes the design and development of an
MC68008-based burglar alarm with particular reference to the software and hardware debugging techniques that were used. The design and debugging was performed without the use of an emulator,
the traditional development tool, and used cross-compilers and
evaluation boards instead. The design process was carefully controlled to allow the gradual integration of the system with one unknown
at a time. The decision process behind the compiler choice, the higher
level software development and testing, software porting, hardware
testing and integration are fully explained.

Design goals

The system under design was an MC68008-based intelligent
burglar alarm which scanned and analysed sensor inputs to screen
out transient and false alarms. The basic hardware consisted of a
processor, a 2k × 8 static RAM, a 32k × 8 EPROM and three octal
latches (74LS373) to provide 16 sensor and data inputs and 8 outputs.

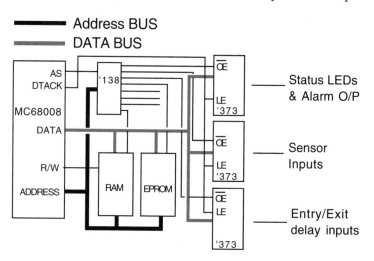

The simplified target hardware

A 74LS138 was used to generate chip selects and output
enables for the memory chips and latches from three higher order
address lines. Three lines were left for future expansion. The sirens etc
were driven via 5 volt gate power MOSFETs. The controlling software
was written in C and controlled the whole timing and response of the
alarm system. Interrupts were not used and the power on reset signal
generated using a CR network and a Schmidt gate.

Development strategy

The normal approach would be to use an in-circuit emulator to debug the software and target hardware, but it was decided at an early stage to develop the system without using an emulator except as a last resort. The reasoning was simple:

- The unit was a replacement for a current analogue system, and the physical dimensions of the case effectively prevented the insertion of an emulation probe. In addition, the case location was very inaccessible.

- The hardware design was a minimum system which relied on the MC68008 signals to generate the asynchronous hand-shakes automatically, e.g. the address strobe is immediately routed back to generate a DTACK signal. This configuration reduces the component count but any erroneous accesses are not recognised. While these timings and techniques are easy to use with a processor, the potential timing delays caused by an emulator could cause problems which are exhibited when real silicon is used.

- The software development was performed in parallel with the hardware development and it was important that the software was tested in a close an environment as possible to a debugged target system early on in the design. While emulators can provide a simple hardware platform, they can have difficulties in coping with power-up tests and other critical functions.

The strategy was finally based on several policies:

- At every stage, only one unknown would be introduced to allow the fast and easy debugging of the system, e.g. software modules were developed and tested on a known working hardware platform, cross-compiled and tested on an evaluation board etc.

- An evaluation board would be used to provide a working target system for the system software debugging. One of the keys to the project was the closeness of this environment to the target system.

- Test modules would be written to test hardware functionality of the target system, and these were tested on the evaluation board.

- The system software would only be integrated on the target board if the test modules executed correctly.

Software development

The first step in the development of the software was to test the logic and basic software design using a workstation. A UNIX workstation was initially used and this allowed the bulk of the software to be generated, debugged and functionally tested within a known working hardware environment, thus keeping with the single unknown strategy. This restricts any new unknown software or hardware to a single component and so makes debugging easier to perform.

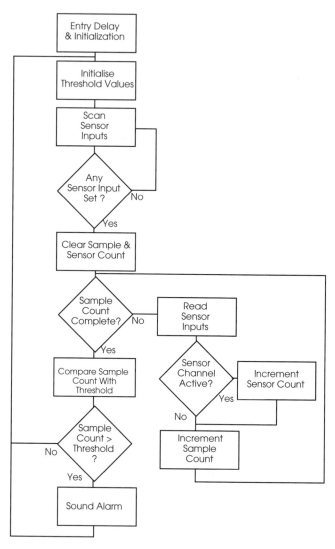

The main software flow diagram

Sensor inputs were simulated using getchar() to obtain values from a keyboard, and by using the multitasking signalling available with a UNIX environment. As a result, the keyboard could be used to input values with the hexadecimal value representing the input port value. Outputs were simulated using a similar technique using printf() to display the information on the screen. Constants for software delays etc were defined using #define C preprocessor statements to allow their easy modification. While the system could not test the software in real time, it does provide functional and logical testing.

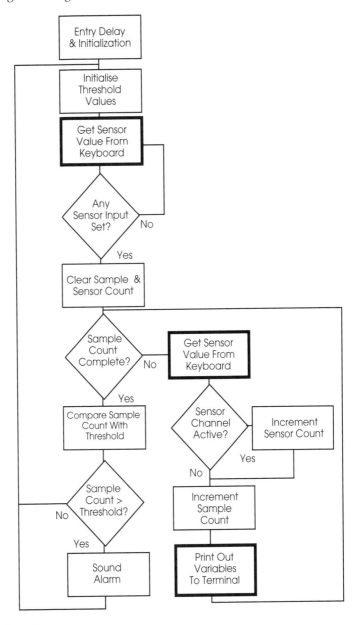

The modified software flow diagram

While it is easy to use the getchar() routine to generate an integer value based on the ASCII value of the key that has been pressed, there are some difficulties. The first problem that was encountered was that the UNIX system uses buffered input for the keyboard. This meant that the return key had to be pressed to signal the UNIX machine to pass the data to the program. This initially caused a problem in that it stopped the polling loop to wait for the keyboard every time it went round the loop. As a result, the I/O simulation was not very good or representative.

In the end, two solutions were tried to improve the simulation. The simplest one was to use the waiting as an aid to testing the software. By having to press the return key, the execution of the polling loop could be brought under the user's control and by printing out the sample and sensor count variables, it was possible to step through each loop individually. By effectively executing discrete polling loops and have the option of simulating a sensor input by pressing a key before pressing return, other factors like the threshold values and the disabling of faulty sensors could be exercised.

The more complex solution was to use the UNIX multitasking and signalling facilities to set-up a second task to generate simple messages to simulate the keyboard input. While this allowed different sensor patterns to be sent to simulate various false triggering sequences without having the chore of having to calculate the key sequences needed from the keyboard, it did not offer much more than that offered by the simple solution.

The actual machine used for this was an OPUS personal mainframe based on a MC88100 RISC processor running at 25 MHz and executing UNIX System V/88. The reasons behind this decision were many:

- It provided an extremely fast system, which dramatically reduced the amount of time spent in compilation and development.

- The ease of transferring software source to other systems for cross-compilation. The Opus system can directly read and write MS-DOS files and switch from the UNIX to PC environment at a keystroke.

- The use of a UNIX-based C compiler, along with other utilities such as lint, was perceived to provide a more standard C source than is offered by other compilers.

This was deemed important to prevent problems when cross-compiling. Typical errors that have been encountered in the past are: byte ordering, variable storage sizes, array and variable initialisation assumed availability of special library routines etc.

Cross-compilation and code generation

Three MC68000 compilation environments were available: the first was a UNIX C compiler running on a VMEbus system, the second was a PC-based cross-compiler supplied with the Motorola MC68020 evaluation board, while a third option of another PC-based cross compiler was a possibility. The criteria for choosing the cross-compilation development were:

- The ease of modifying run time libraries to execute standalone on the MC68020 evaluation board and, finally the target system.

- The quality of code produced.

The second option was chosen primarily for the ease with which the runtime libraries could be modified. As standard, full runtime support was available for the evaluation board, and these modules, including the all-important initialisation routines, were supplied in assembler source and are very easy to modify. Although the code quality was not as good as the other options, it was adequate for the design and the advantage of immediate support for the software testing on the evaluation board more than compensated. This support fitted in well with the concept of introducing only a single unknown.

If the software can be tested in as close an environment as possible to the target, any difficulties should lie with the hardware design. With this design, the only differences between the target system configuration and the evaluation board is a different memory map. Therefore, by testing the software on an evaluation board, one of the unknowns can be removed when the software is integrated with the target system. If the runtime libraries are already available, this further reduces the unknowns to that of just the system software.

The C source was then transferred to a PC for cross-compilation. The target system was a Flight MC68020 evaluation board which provides a known working MC68xxx environment and an on board debugger. The code was cross-compiled, downloaded via a serial link and tested again.

The testing was done in two stages: the first simply ran the software that had been developed using the OPUS system without modifying the code. This involved using the built-in getchar() system calls and so on within the Flight board debugger. It is at this point that any differences between the C compilers would be found such as array initialisation, bit and byte alignment etc.. It is these differences that can prevent software being ported from one system to another. These tests provided further confidence in the software and its functionality before further modification.

The second stage was to replace these calls by a pointer to a memory location. This value would be changed through the use of the onboard debugger. The evaluation board abort button is pressed to stop the program execution and invoke the debugger. The corresponding memory location is modified and the program execution restarted. At this point, the software is virtually running as expected on the target system. All that remains to do is to take into account the target hardware memory map and initialisation.

Porting to the final target system

The next stage involved porting the software to the final target configuration. These routines allocate the program and stack memory, initialise pointers etc. and define the sensor and display locations within the memory map. All diagnostic I/O calls were removed. The cross-compiler used supplies a startup assembly file which performs these tasks. This file was modified and the code recompiled all ready for testing.

Generation of test modules

Although the target hardware design was relatively simple, it was thought prudent to generate some test modules which would exercise the memory and indicate the success by lighting simple status LEDs. Although much simpler than the controlling software, these go-nogo tests were developed using the same approach: written and tested on the evaluation board, changed to reflect the target configuration and then blown into EPROM.

The aim of these tests was to prove that the hardware functioned correctly: the critical operations that were tested included power-up reset and initialisation, reading and writing to the I/O functions, and exercising the memory.

These routines were written in assembler and initially tested using the Microtec Xray debugger and simulator before downloading to the Flight board for final testing.

Target hardware testing

After checking the wiring, the test module EPROM was installed and the target powered up. Either the system would work or not. Fortunately, it did! With the hardware capable of accessing memory, reading and writing to the I/O ports, the next stage was to install the final software.

While the system software logically functioned, there were some timing problems associated with the software timing loops which controlled the sample window, entry/exit time delays and alarm duration. These errors were introduced as a direct result of the software methodology chosen: the delay values would obviously change depending on the hardware environment, and while the values were defined using #define preprocessor statements and were adjusted to reflect the processing power available, they were inaccurate. To solve this problem, some additional test modules were written, and by using trial and error, the correct values were derived. The system software was modified and installed.

Future techniques

The software loop problem could have been solved if an MC68000 software simulator had been used to execute, test and time the relevant software loops. This would have saved a day's work.

If a hardware simulator had been available, it could have tested the hardware design and provided additional confidence that it was going to work.

Relevance to more complex designs

The example described is relatively simple but many of the techniques used are extremely relevant to more complex designs. The fundamental aim of a test and development methodology, which restricts the introduction of untested software or hardware to a single item, is a good practice and of benefit to the design and implementation of any system, even those that use emulation early on in the design cycle.

The use of evaluation boards or even standalone VME or Multibus II boards can be of benefit for complex designs. The amount of benefit is dependent of the closeness of the evaluation board hardware to the target system. If this design had needed to use two serial ports, timers and parallel I/O, it is likely that an emulator would still not have been used provided a ready built board was available which used the same peripheral devices as the target hardware. The low level software drivers for the peripherals could be tested on the evaluation board and these incorporated into the target test modules for hardware testing.

There are times, however, when a design must be tested and an emulator is simply not available. This scenario occurs when designing with a processor at a very early stage of product life. There is inevitably a delay between the appearance of working silicon and instrumentation support. During this period, similar techniques to those described are used and a logic analyser used instead of a emulator to provide instrumentation support, in case of problems. If the processor is a new generation within an existing family, previous members can be used to provide an interim target for some software testing. If the design involves a completely new processor, similar techniques can be applied, except at some point untested software must be run on untested hardware.

It is to prevent this integration of two unknowns that simulation software to either simulate and test software, hardware or both can play a critical part in the development process.

The need for emulation

Even using the described techniques, it cannot be stated that there will never be a need for additional help. There will be times when instrumentation, such as emulation and logic analysis, is necessary to resolve problems within a design quickly. Timing and intermittent problems cannot be easily solved without access to further information about the processor and other system signals. Even so, the recognition of a potential problem source such as a specific software module or hardware allows a more constructive use and a speedier resolution. The adoption of a methodical design approach and the use of ready built boards as test vehicles may, at best, remove the need for emulation and, at worst, reduce the amount of time debugging the system.

Digital echo unit

This design example follows the construction of a digital echo unit to provide echo and reverb effects.

With sound samples digitally recorded, it is possible to use digital signal processing techniques to create far better and more flexible effects units (or sound processors, as they are more commonly called). Such units comprise a fast digital signal processor with A to D and D to A converters and large amounts of memory. An analogue signal is sent into the processor, converted into the digital domain, processed using software running on the processor to create filters, delay, reverb and other effects before being converted back into an analogue signal and being sent out.

They can be completely software based, which provides a lot of flexibility, or they can be preprogrammed. They can take in analogue or, in some cases, digital data, and feed it back into other units or directly into an amplifier or audio mixing desk, just like any other instrument.

Creating echo and reverb

Analogue echo and reverb units usually rely on an electrome-chanical method of delaying an audio signal to create reverberation or echo. The WEM Copycat used a tape loop and a set of tape heads to record the signal onto tape and then read it from the three or more tape heads to provide three delayed copies of the signal.

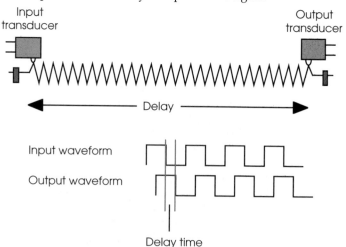

Spring line delay

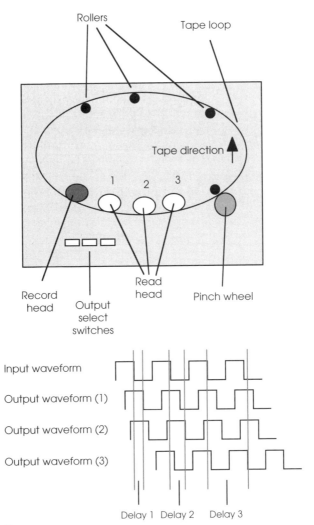

A tape loop-based delay unit

The delay was a function of the tape speed and the distance between the recording and read tape heads. This provides a delay of up to 1 second. Spring line delays used a transducer to send the audio signal mechanically down a taut spring where the delayed signal would be picked up again by another transducer.

Bucket brigade devices have also been used to create a purely electronic delay. These devices take an analogue signal and pass it from one cell to another using a clock. The technique is similar to passing a bucket of water by hand down a line of men. Like the line of men, where some water is inevitably lost, the analogue signal degrades — but it is good enough to achieve some good effects.

With a digitised analogue signal, creating delayed copies is easy. The samples can be stored in memory in a buffer and later retrieved. The advantage this offers is that the delayed sample is an exact copy of the original sound and, unlike the techniques previously described, has not degraded in quality or had tape noise introduced. The number of delayed copies is dependent on the number of buffers and hence the amount of memory that is available. This ability, coupled with a signal processor allows far more accurate and natural echoes and reverb to be created.

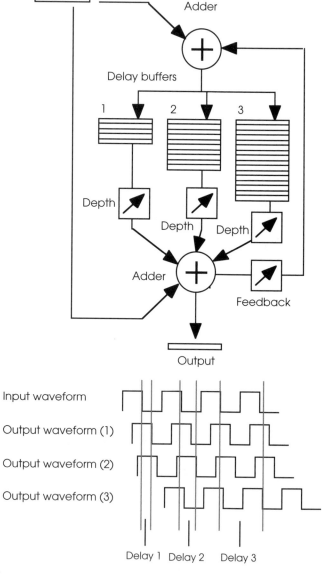

A digital echo/reverb unit

The problem with many analogue echo and reverb units is that they simplify the actual reverb and echo. In natural conditions, such as a large concert hall, there are many delay sources as the sound bounces around and this cannot be reproduced with only two or three voices which are independently mixed together with a bit of feedback. The advantage of the digital approach is that as many delays can be created as required and the signal processor can combine and fade the different sources as needed to reproduce the environment required.

The block diagram shows how such a digital unit can be constructed. The design uses three buffers to create three separate delays. These buffers are initially set to zero and are FIFOs — first in, first out —thus the first sample to be placed at the top of each buffer appears at the bottom at different times and is delayed by the number of samples that each buffer holds. The smaller the buffer, the smaller the delay. The outputs of the three buffers are all individually reduced in size according to the depth required or the prominence that the delayed sound has. A large value gives a very echoey effect, similar to that of a large room or hall. A small depth reduces it. The delayed samples are combined by the adder with the original sample — hence the necessity to clear the buffer initially to ensure that random values, which adds noise, do not get added before the first sample appears — to create the final effect. A feedback loop takes some of the output signal, as determined by the feedback control, and combines it with the original sample as it is stored in the buffers. This effectively controls the decay of the delayed sounds and creates a more natural effect.

This type of circuit can become more sophisticated by adapting the depth with time and having separate independent feedback loops and so on. This circuit can also be the basis of other effects such as chorus, phasing and flanging where the delayed signal is constantly delayed but varies. This can be done by altering the timing of the sample storage into the buffers.

Design requirements

The design requirements for the echo unit are as follows:

- It must provide storage for at least one second on all its channels.
- It must provide control over the echo length and depth.
- It must take analogue signals in and provide analogue signals out.
- The audio quality must be good with a 20 kHz bandwidth.

Designing the codecs

The first decision concerns the A to D and D to A codec design. Many lower specification units use 8 bit A to D and D to A units to digitise and convert the delayed analogue signal. This signal does not need to be such good quality as the original and using an 8 bit resolution converter saves on cost and reduces the amount of memory needed to store the delayed signal. Such systems normally add the delayed signal in the analogue domain and this helps to cover any quality degradation.

With this design, the quality requirement precludes the use of 8 bit converters and effectively dictates that a higher quality codec is used. With the advent of Compact Disc, there are now plenty of high quality audio codecs available with sample sizes of 12 or more bits. A top end device would use 16 bit conversion and this would fit nicely with 16 bit memory. This is also the sample size used with Compact Disc.

The next consideration is the conversion rate. To achieve a bandwidth of 20 kHz, a conversion rate of 40 kHz is needed. This has several knock-on effects: it determines the number of samples needed to store one second of digital audio and hence the amount of memory. It also defines the timing that the system must adhere to remove any sampling errors. The processor must be able to receive the digitised audio, store it and copy it as necessary, retrieve the output samples, combine them and convert them to the analogue signals every 25 µs.

Designing the memory structures

In examining the codec design, some of the memory requirements have already started to appear. The first requirement is the memory storage for the digital samples. For a single channel of delay where only a single delayed audio signal is combined with the original signal, the memory storage is the sample size multiplied by the sample rate and the total storage time taken. For a 16 bit sample and a 40 kHz rate, 80000 bytes of storage needed. Rounding up, this is equivalent to just over 78 kbytes of storage (a kbyte of memory is 1024 bytes and not 1000).

This memory needs to be organised as a by 16 structure which means that the final design will need 40 k by 16 words of memory per second of audio. For a system with three delayed audio sources, this is about 120 k words which works out very nicely at two 128k by 8 RAM chips. The spare 8 kbytes in each RAM chip can be used by the supervisor software that will run on the control processor.

Now that the amount of memory is known, then the memory type and access speed can be worked out. DRAM is applicable in this case but requires refresh circuitry and because it is very high density may not be cost effective. If 16 Mb DRAM is used then with a by 16 organisation, a single chip would provide 1 Mbyte of data storage which is far too much for this application. The other potential problem is the effect of refresh cycles which would potentially introduce sampling errors. This means that static RAM is probably the best solution.

To meet the 25 µs cycle time which includes a minimum of a data read and a data write, this means that the overall access time must be significantly less than half of the cycle time, i.e. less than 12.5 μs. This means that almost any memory is capable of performing this function.

In addition, some form of non-volatile memory is needed to contain the control software. This would normally be stored in an EPROM. However, the EPROM access times are not good and therefore may not be suitable for running the software directly. If the control program is small enough, then it could be transferred from the EPROM to the FSRAM and executed from there.

The software design

The software design is relatively simple and treats the process as a pipeline. While the A to D is converting the next sample, the previous sample is taken and stored in memory using a circular buffer to get the overall delay effect. The next sample for D to A conversion and output is retrieved from the buffer and sent to the converter. The circular buffer pointers are then updated, including checking for the end of the buffer.

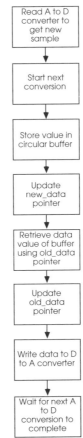

The basic pipeline flow for the software

This sequence is repeated every 25 µs. While the processor is not performing this task, it can check and maintain the user controls. As stated previously, circular buffers are used to hold the digitised data. A buffer is used with two pointers: one points to the next storage location for the incoming data and a second pointer is used to locate the delayed data. The next two diagrams show how this works. Each sample is stored consecutively in memory and the two pointers are separated by a constant value which is equivalent to the number of samples delay that is required. In the example shown, this is 16 samples. This difference is maintained so that when a new sample is inserted, the corresponding old value is removed as well and then both pointers are updated.

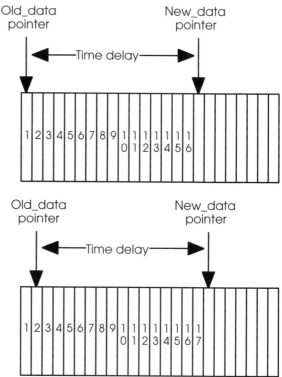

Using a circular buffer to store and retrieve data

When either pointer reaches the end of the data block, its value is changed to point to the next location. In the example shown, the New_data pointer is reset to point at the first location in the buffer which held the first sample. This sample is no longer needed and its value can be overwritten. By changing the difference between the two pointers, the time delay can be changed. In practice, the pointers are simply memory addresses and every time they are updated, they should be checked and if necessary reset to the beginning of the table. This form of addressing is known as modulo addressing and some

DSP processors support it directly and therefore do not need to check the address.

When using these structures, it is important to ensure that all values are set initially to zero so that the delayed signal is not random noise when the system first starts up. The delayed signal will not be valid until the buffer has filled. In the examples shown, this would be 16 samples. Until this point, the delayed signal will be made from the random contents of the buffer. By clearing these values to zero, silence is effectively output and no noise is heard until the correct delayed signal.

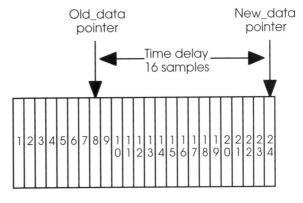

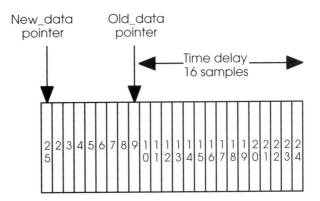

Implementing modulo addressing

Multiple delays

With a multiple source system, the basic software design remains intact except that the converted data is copied into several delay buffers and the outputs from these buffers are combined before the end result is converted into the analogue signal.

There are several ways of setting this up. The first is to use multiple buffers and copy each new value into each buffer. Each buffer then supplies its own delayed output which can be combined to create the final effect. A more memory efficient system is to use a single buffer but add additional old_data pointers with different time delays to create the different delay length outputs.

The overhead in doing this is small. There is the maintenance of the pointers to be done and the combination of the delay values to create the final output for the D to A converter. This can be quite complex depending on the level of sophistication needed.

Digital or analogue adding

There are some options depending on the processing power available. With a real echo or reverb, the delayed signals need to be gradually attenuated as the signals die away and therefore, the delayed signal must be attenuated. This can be done either digitally or in the analogue domain. With a single source, the analogue implementation is easy. The delayed signal is converted and an analogue mixer is used to attenuate and combine the delayed signal with the original to create the reverb or echo effect. An analogue feedback bath can also be created.

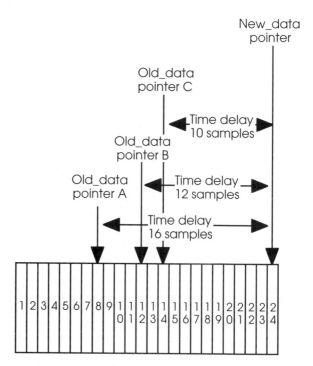

Using a single buffer with multiple pointers to create multiple delays

The multiple delayed source design can use this same analogue method but requires a separate D to A converter for each delayed signal. This can be quite expensive. Instead, the processor can add the signals together, along with attenuation factors to create a combined delay signal that can be sent to the D to A converter for combination with the original analogue signal. It is therefore possible to perform all this, including the combination with the original signal in the digital domain, and then simply output the end value to the D to A converter. With this version, the attenuation does not need to be constant and can be virtually any type of curve.

The disadvantage is the computation that is needed. The arithmetic that is required is saturation arithmetic which is a little more than simply adding two values together. This is needed to ensure that the combined value only provides a peak value and not cause an overflow error. In addition, all the calculations must be done within 25 µs to meet the sampling rate criteria and this can be pushing the design a little with many general-purpose processors.

Microprocessor selection

The choice of microprocessor is dependent on several factors. It must have an address range of greater than 64 kbytes and have a 16 bit data path. It must be capable of performing 16 bit arithmetic and thus this effectively rules out 8 bit microprocessors and microcontrollers.

In terms of architecture, multiple address pointers that auto-increment would make the circular buffer implementations very efficient and therefore some like a RISC processor or a fast MC68000 would be suitable. Other architectures can certainly due the job but their additional overhead may reduce their ability to perform all the processing within the 25 µs window that there is. One way of finding this out is to create some test code and run it on a simulator or emulator, for example, to find out how many clocks it does take to execute these key routines.

A low cost DSP processor is also quite attractive in this type of application, especially if it supports modulo addressing and saturation arithmetic.

The overall system design

The basic design for the system uses a hardware timer to generate a periodic interrupt every 25 µs. The associated interrupt service routine is where the data from the A to D converter is read and stored, the next conversion started and the delayed data taken from the buffer and combined. The pointers are updated before returning from the service routine. In this way, the sampling is done on a regular basis and is given a higher priority than the background processing.

While the processor is not servicing the interrupt, it stays in a forever loop, polling the user interface for parameters and commands. The delay times are changed by manipulating the pointers. It is possible to do this by changing the sampling rate instead but the audio quality does not stay constant.

The system initialises by clearing the RAM to zero and using some of it to hold the program code which is copied from EPROM. If a battery backed SRAM is used instead, then used defined parameters and settings could be stored here as well and retained when the system is switched off.

Index

Symbols